LONE STAR PLANET

A Planet for Texans

H. Beam Piper

Lone Star Planet

Table of Contents

Chapter I
Chapter II
Chapter III
Chapter IV
Chapter V
Chapter VI
Chapter VII
Chapter VIII
Chapter IX
Chapter X
Chapter XI

Chapter I

They started giving me the business as soon as I came through the door into the Secretary's outer office.

There was Ethel K'wang–Li, the Secretary's receptionist, at her desk. There was Courtlant Staynes, the assistant secretary to the Undersecretary for Economic Penetration, and Norman Gazarin, from Protocol, and Toby Lawder, from Humanoid Peoples' Affairs, and Raoul Chavier, and Hans Mannteufel, and Olga Reznik.

It was a wonder there weren't more of them watching the condemned man's march to the gibbet: the word that the Secretary had called me in must have gotten all over the Department since the offices had opened.

"Ah, Mr. Machiavelli, I presume," Ethel kicked off.

"Machiavelli, Junior." Olga picked up the ball. "At least, that's the way he signs it."

"God's gift to the Consular Service, and the Consular Service's gift to Policy Planning," Gazarin added.

"Take it easy, folks. These Hooligan Diplomats would as soon shoot you as look at you," Mannteufel warned.

"Be sure and tell the Secretary that your friends all want important posts in the Galactic Empire." Olga again.

"Well, I'm glad some of you could read it," I fired back. "Maybe even a few of you understood what it was all about."

"Don't worry, Silk," Gazarin told me. "Secretary Ghopal understands what it was all about. All too well, you'll find."

A buzzer sounded gently on Ethel K'wang–Li's desk. She snatched up the handphone and whispered into it. A deathly silence filled the room while she listened, whispered some more, then hung it up.

They were all staring at me.

"Secretary Ghopal is ready to see Mr. Stephen Silk," she said. "This way, please."

As I started across the room, Staynes began drumming on the top of the desk with his fingers, the slow reiterated rhythm to which a man marches to a military execution.

"A cigarette?" Lawder inquired tonelessly. "A glass of rum?"

There were three men in the Secretary of State's private office. Ghopal Singh, the Secretary, dark–faced, gray–haired, slender and elegant, meeting me halfway to his desk. Another slender man, in black, with a silver–threaded, black neck–scarf: Rudolf Klüng, the Secretary of the Department of Aggression.

And a huge, gross–bodied man with a fat baby–face and opaque black eyes.

When I saw him, I really began to get frightened.

The fat man was Natalenko, the Security Coördinator.

"Good morning, Mister Silk," Secretary Ghopal greeted me, his hand extended. "Gentlemen, Mr. Stephen Silk, about whom we were speaking. This way, Mr. Silk, if you please."

There was a low coffee–table at the rear of the office, and four easy chairs around it. On the round brass table–top were cups and saucers, a coffee urn, cigarettes—and a copy of the current issue of the *Galactic Statesmen's Journal*, open at an article entitled *Probable Future Courses of Solar League Diplomacy*, by somebody who had signed himself Machiavelli, Jr.

I was beginning to wish that the pseudonymous Machiavelli, Jr. had never been born, or, at least, had stayed on Theta Virgo IV and been a wineberry planter as his father had wanted him to be.

As I sat down and accepted a cup of coffee, I avoided looking at the periodical. They were probably going to hang it around my neck before they shoved me out of the airlock.

"Mr. Silk is, as you know, in our Consular Service," Ghopal was saying to the others. "Back on Luna on rotation, doing something in Mr. Halvord's section. He is the gentleman who did such a splendid job for us on Assha—Gamma Norma III.

"And, as he has just demonstrated," he added, gesturing toward the *Statesman's Journal* on the Benares–work table, "he is a student both of the diplomacy of the past and the implications of our present policies."

"A bit frank," Klüng commented dubiously.

"But judicious," Natalenko squeaked, in the high eunuchoid voice that came so incongruously from his bulk. "He aired his singularly accurate predictions in a periodical that doesn't have a circulation of more than a thousand copies outside his own department. And I don't think the public's semantic reactions to the terminology of imperialism is as bad as you imagine. They seem quite satisfied, now, with the change in the title of your department, from Defense to Aggression."

"Well, we've gone into that, gentlemen," Ghopal said. "If the article really makes trouble for us, we can always disavow it. There's no censorship of the *Journal*. And Mr. Silk won't be around to draw fire on us."

Here it comes, I thought.

"That sounds pretty ominous, doesn't it, Mr. Silk?" Natalenko tittered happily, like a ten–year–old who has just found a new beetle to pull the legs out of.

"It's really not as bad as it sounds, Mr. Silk," Ghopal hastened to reassure me. "We are going to have to banish you for a while, but I daresay that won't be so bad. The social life here on Luna has probably begun to pall, anyhow. So we're sending you to Capella IV."

"Capella IV," I repeated, trying to remember something about it. Capella was a GO–type, like Sol; that wouldn't be so bad.

"New Texas," Klüng helped me out.

Oh, God, no! I thought.

"It happens that we need somebody of your sort on that planet, Mr. Silk," Ghopal said. "Some of the trouble is in my department and some of it is in Mr. Klüng's; for that reason, perhaps it would be better if Coördinator Natalenko explained it to you."

"You know, I assume, our chief interest in New Texas?" Natalenko asked.

"I had some of it for breakfast, sir," I replied. "Supercow."

Natalenko tittered again. "Yes, New Texas is the butcher shop of the galaxy. In more ways than one, I'm afraid you'll find. They just butchered one of our people there a short while ago. Our Ambassador, in fact."

That would be Silas Cumshaw, and this was the first I'd heard about it.

I asked when it had happened.

"A couple of months ago. We just heard about it last evening, when the news came in on a freighter from there. Which serves to point up something you stressed in your article—the difficulties of trying to run a centralized democratic government on a galactic scale. But we have another interest, which may be even more urgent than our need for New Texan meat. You've heard, of course, of the z'Srauff."

That was a statement, not a question; Natalenko wasn't trying to insult me. I knew who the z'Srauff were; I'd run into them, here and there. One of the extra–solar intelligent humanoid races, who seemed to have been evolved from canine or canine–like ancestors, instead of primates. Most of them could speak Basic English, but I never saw one who would admit to understanding more of our language than the 850–word Basic vocabulary. They occupied a half–dozen planets in a small star–cluster about forty light–years beyond the Capella system. They had developed normal–space reaction–drive ships before we came into contact with them, and they had quickly picked up the hyperspace–drive from us back in those days when the Solar League was still playing Missionaries of Progress and trying to run a galaxy–wide Point–Four program.

In the past century, it had become almost impossible for anybody to get into their star–group, although z'Srauff ships were orbiting in on every planet that the League had settled or controlled. There were z'Srauff traders and small merchants all over the galaxy, and you almost never saw one of them without a camera. Their little meteor-mining boats were everywhere, and all of them carried more of the most modern radar and astrogational equipment than a meteor-miner's lifetime earnings would pay for.

I also knew that they were one of the chief causes of ulcers and premature gray hair at the League capital on Luna. I'd done a little reading on pre-spaceflight Terran history; I had been impressed by the parallel between the present situation and one which had culminated, two and a half centuries before, on the morning of 7 December, 1941.

"What," Natalenko inquired, "do you think Machiavelli, Junior would do about the z'Srauff?"

"We have a Department of Aggression," I replied. "Its mottoes are, 'Stop trouble before it starts,' and, 'If we have to fight, let's do it on the other fellow's real estate.' But this situation is just a little too delicate for literal application of those principles. An unprovoked attack on the z'Srauff would set every other non-human race in the galaxy against us.... Would an attack by the z'Srauff on New Texas constitute just provocation?"

"It might. New Texas is an independent planet. Its people are descendants of emigrants from Terra who wanted to get away from the rule of the Solar League. We've been trying for half a century to persuade the New Texan government to join the League. We need their planet, for both strategic and commercial reasons. With the z'Srauff for neighbors, they need us as much at least as we need them. The problem is to make them understand that."

I nodded again. "And an attack by the z'Srauff would do that, too, sir," I said.

Natalenko tittered again "You see, gentlemen! Our Mr. Silk picks things up very handily, doesn't he?" He turned to Secretary of State Ghopal. "You take it from there," he invited.

Ghopal Singh smiled benignly. "Well, that's it, Stephen," he said. "We need a man on New Texas who can get things done. Three things, to be exact.

"First, find out why poor Mr. Cumshaw was murdered, and what can be done about it to maintain our prestige without alienating the New Texans.

"Second, bring the government and people of New Texas to a realization that they need the Solar League as much as we need them.

"And, third, forestall or expose the plans for the z'Srauff invasion of New Texas."

Is that all, now? I thought. *He doesn't want a diplomat; he wants a magician.*

"And what," I asked, "will my official position be on New Texas, sir? Or will I have one, of any sort?"

"Oh, yes, indeed, Mr. Silk. Your official position will be that of Ambassador Plenipotentiary and Envoy Extraordinary. That, I believe, is the only vacancy which exists in the Diplomatic Service on that planet."

At Dumbarton Oaks Diplomatic Academy, they haze the freshmen by making them sit on a one–legged stool and balance a teacup and saucer on one knee while the upper classmen pelt them with ping–pong balls. Whoever invented that and the other similar forms of hazing was one of the great geniuses of the Service. So I sipped my coffee, set down the cup, took a puff from my cigarette, then said:

"I am indeed deeply honored, Mr. Secretary. I trust I needn't go into any assurances that I will do everything possible to justify your trust in me."

"I believe he will, Mr. Secretary," Natalenko piped, in a manner that chilled my blood.

"Yes, I believe so," Ghopal Singh said. "Now, Mr. Ambassador, there's a liner in orbit two thousand miles off Luna, which has been held from blasting off for the last eight hours, waiting for you. Don't bother packing more than a few things; you can get everything you'll need aboard, or at New Austin, the planetary capital. We have a man whom Coördinator Natalenko has secured for us, a native New Texan, Hoddy Ringo by name. He'll act as your personal secretary. He's aboard the ship now. You'll have to hurry, I'm afraid.... Well, *bon voyage*, Mr. Ambassador."

Chapter II

The death–watch outside had grown to about fifteen or twenty. They were all waiting in happy anticipation as I came out of the Secretary's office.

"What did he do to you, Silk?" Courtlant Staynes asked, amusedly.

"Demoted me. Kicked me off the Hooligan Diplomats," I said glumly.

"Demoted you from the Consular Service?" Staynes asked scornfully. "Impossible!"

"Yes. He demoted me to the Cookie Pushers. Clear down to Ambassador."

They got a terrific laugh. I went out, wondering what sort of noises they'd make, the next morning, when the appointments sheet was posted.

I gathered a few things together, mostly small personal items, and all the microfilms that I could find on New Texas, then got aboard the Space Navy cutter that was waiting to take me to the ship. It was a four–hour trip and I put in the time going over my hastily–assembled microfilm library and using a stenophone to dictate a reading list for the spacetrip.

As I rolled up the stenophone–tape, I wondered what sort of secretary they had given me; and, in passing, why Natalenko's department had furnished him.

Hoddy Ringo....

Queer name, but in a galactic civilization, you find all sorts of names and all sorts of people bearing them, so I was prepared for anything.

And I found it.

I found him standing with the ship's captain, inside the airlock, when I boarded the big, spherical space–liner. A tubby little man, with shoulders and arms he had never developed doing secretarial work, and a good–natured, not particularly intelligent face.

See the happy moron, he doesn't give a damn, I thought.

Then I took a second look at him. He might be happy, but he wasn't a moron. He just looked like one. Natalenko's people often did, as one of their professional assets.

I also noticed that he had a bulge under his left armpit the size of an eleven–mm army automatic.

He was, I'd been told, a native of New Texas. I gathered, after talking with him for a while, that he had been away from his home planet for over five years, was glad to be going back, and especially glad that he was going back under the protection of Solar League diplomatic immunity.

In fact, I rather got the impression that, without such protection, he wouldn't have been going back at all.

I made another discovery. My personal secretary, it seemed, couldn't read stenotype. I found that out when I gave him the tape I'd dictated aboard the cutter, to transcribe for me.

"Gosh, boss. I can't make anything out of this stuff," he confessed, looking at the combination shorthand–Braille that my voice had put onto the tape.

"Well, then, put it in a player and transcribe it by ear," I told him.

He didn't seem to realize that that could be done.

"How did you come to be sent as my secretary, if you can't do secretarial work?" I wanted to know.

He got out a bag of tobacco and a book of papers and began rolling a cigarette, with one hand.

"Why, shucks, boss, nobody seemed to think I'd have to do this kinda work," he said. "I was just sent along to show you the way around New Texas, and see you don't get inta no trouble."

He got his handmade cigarette drawing, and hitched the strap that went across his back and looped under his right arm. "A guy that don't know the way around can get inta a lotta trouble on New Texas. If you call gettin' killed trouble."

So he was a bodyguard ... and I wondered what else he was. One thing, it would take him forty–two years to send a radio message back to Luna, and I could keep track of any other messages he sent, in letters or on tape, by ships. In the end, I transcribed my own tape, and settled down to laying out my three weeks' study–course on my new post.

I found, however, that the whole thing could be learned in a few hours. The rest of what I had was duplication, some of it contradictory, and it all boiled down to this:

Capella IV had been settled during the first wave of extrasolar colonization, after the Fourth World—or First Interplanetary—War. Some time around 2100. The settlers had come from a place in North America called Texas, one of the old United States. They had a lengthy history—independent republic, admission to the United States, secession from the United States, reconquest by the United States, and general intransigence under the United States, the United Nations and the Solar League. When the laws of non–Einsteinian physics were discovered and the hyperspace–drive was developed, practically the entire population of Texas had taken to space to find a new home and independence from everybody.

They had found Capella IV, a Terra–type planet, with a slightly higher mean temperature, a lower mass and lower gravitational field, about one–quarter water and three–quarters land–surface, at a stage of evolutionary development approximately that of Terra during the late Pliocene. They also found supercow, a big mammal looking like the unsuccessful attempt of a hippopotamus to impersonate a dachshund and about the size of a nuclear–steam locomotive. On New Texas' plains, there were billions of them; their meat was fit for the gods of Olympus. So New Texas had become the meat–supplier to the galaxy.

There was very little in any of the microfilm–books about the politics of New Texas and such as it was, it was very scornful. There were such expressions as 'anarchy tempered by assassination,' and 'grotesque parody of democracy.'

There would, I assumed, be more exact information in the material which had been shoved into my hand just before boarding the cutter from Luna, in a package labeled *TOP SECRET: TO BE OPENED ONLY IN SPACE, AFTER THE FIRST HYPERJUMP*. There was also a big trunk that had been placed in my suite, sealed and bearing the same instructions.

I got Hoddy out of the suite as soon as the ship had passed out of the normal space–time continuum, locked the door of my cabin and opened the parcel.

It contained only two loose–leaf notebooks, both labeled with the Solar League and Department seals, both adorned with the customary bloodthirsty threats against the unauthorized and the indiscreet. They were numbered *ONE* and *TWO*.

ONE contained four pages. On the first, I read:

FINAL MESSAGE OF THE FIRST SOLAR LEAGUE AMBASSADOR TO NEW TEXAS ANDREW JACKSON HICKOCK

I agree with none of the so–called information about this planet on file with the State Department on Luna. The people of New Texas are certainly not uncouth barbarians. Their manners and customs, while lively and unconventional, are most charming. Their dress is graceful and practical, not grotesque; their soft speech is pleasing to the ear. Their flag is the original flag of the Republic of Texas; it is definitely not a barbaric travesty of our own emblem. And the underlying premises of their political system should, as far as possible, be incorporated into the organization of the Solar League. Here politics is an exciting and exacting game, in which only the true representative of all the people can survive.

DEPARTMENT ADDENDUM

After five years on New Texas, Andrew Jackson Hickock resigned, married a daughter of a local rancher and became a naturalized citizen of that planet. He is still active in politics there, often in opposition to Solar League policies.

That didn't sound like too bad an advertisement for the planet. I was even feeling cheerful when I turned to the next page, and:

FINAL MESSAGE OF THE SECOND SOLAR LEAGUE AMBASSADOR TO NEW TEXAS CYRIL GODWINSON

Yes and no; perhaps and perhaps not; pardon me; I agree with everything you say. Yes and no; perhaps and perhaps not; pardon me; I agree…

DEPARTMENT ADDENDUM

After seven years on New Texas, Ambassador Godwinson was recalled; adjudged hopelessly insane.

And then:

FINAL MESSAGE OF THE THIRD SOLAR LEAGUE AMBASSADOR TO NEW TEXAS R. F. GULLIS

I find it very pleasant to inform you that when you are reading this, I will be dead.

DEPARTMENT ADDENDUM

Committed suicide after six months on New Texas.

I turned to the last page cautiously, found:

FINAL MESSAGE OF THE FOURTH SOLAR LEAGUE AMBASSADOR TO NEW TEXAS SILAS CUMSHAW

I came to this planet ten years ago as a man of pronounced and outspoken convictions. I have managed to keep myself alive here by becoming an inoffensive nonentity. If I continue in this course, it will be only at the cost of my self–respect. Beginning tonight, I am going to state and maintain positive opinions on the relation between this planet and the Solar League.

DEPARTMENT ADDENDUM

Murdered at the home of Andrew J. Hickcock. (see p. 1.)

And that was the end of the first notebook. Nice, cheerful reading; complete, solid briefing.

I was, frankly, almost afraid to open the second notebook. I hefted it cautiously at first, saw that it contained only about as many pages as the first and that those pages were sealed with a band around them.

I took a quick peek, read the words on the band:

Before reading, open the sealed trunk which has been included with your luggage.

So I laid aside the book and dragged out the sealed trunk, hesitated, then opened it.

Nothing shocked me more than to find the trunk ... full of clothes.

There were four pairs of trousers, light blue, dark blue, gray and black, with wide cuffs at the bottoms. There were six or eight shirts, their colors running the entire spectrum in the most violent shades. There were a couple of vests. There were two pairs of short boots with high heels and fancy leather–working, and a couple of hats with four–inch brims.

And there was a wide leather belt, practically a leather corset.

I stared at the belt, wondering if I was really seeing what was in front of me.

Attached to the belt were a pair of pistols in right– and left–hand holsters. The pistols were seven–mm Krupp–Tatta Ultraspeed automatics, and the holsters were the spring–ejection, quick–draw holsters which were the secret of the State Department Special Services.

This must be a mistake, I thought. *I'm an Ambassador now and Ambassadors never carry weapons.*

The sanctity of an Ambassador's person not only made the carrying of weapons unnecessary, so that an armed Ambassador was a contradiction of diplomatic terms, but it would be an outrageous insult to the nation to which he had been accredited.

Like taking a poison–taster to a friendly dinner.

Maybe I was supposed to give the belt and the holsters to Hoddy Ringo....

So I tore the sealed band off the second notebook and read through it.

I was to wear the local costume on New Texas. That was something unusual; even in the Hooligan Diplomats, we leaned over backward in wearing Terran costume to distinguish ourselves from the people among whom we worked.

I was further advised to start wearing the high boots immediately, on shipboard, to accustom myself to the heels. These, I was informed, were traditional. They had served a useful purpose, in the early days on Terran Texas, when all travel had been on horseback. On horseless and mechanized New Texas, they were a useless but venerated part of the cultural heritage.

There were bits of advice about the hat, and the trousers, which for some obscure reason were known as Levis. And I was informed, as an order, that I was to wear the belt and the pistols at all times outside the Embassy itself.

That was all of the second notebook.

The two notebooks, plus my conversation with Ghopal, Klüng and Natalenko, completed my briefing for my new post.

I slid off my shoes and pulled on a pair of boots. They fitted perfectly. Evidently I had been tapped for this job as soon as word of Silas Cumshaw's death had reached Luna and there must have been some fantastic hurrying to get my outfit ready.

I didn't like that any too well, and I liked the order to carry the pistols even less. Not that I had any objection to carrying weapons, *per se*: I had been born and raised on Theta Virgo IV, where the children aren't allowed outside the house unattended until they've learned to shoot.

But I did have strenuous objections to being sent, virtually ignorant of local customs, on a mission where I was ordered to commit deliberate provocation of the local government, immediately on the heels of my predecessor's violent death.

The author of *Probable Future Courses of Solar League Diplomacy* had recommended the use of provocation to justify conquest. If the New Texans murdered two Solar League Ambassadors in a row, nobody would blame the League for moving in with a space–fleet and an army....

I was beginning to understand how Doctor Guillotin must have felt while his neck was being shoved into his own invention.

I looked again at the notebooks, each marked in red: *Familiarize yourself with contents and burn or disintegrate.*

I'd have to do that, of course. There were a few non–humans and a lot of non–League people aboard this ship. I couldn't let any of them find out what we considered a full briefing for a new Ambassador.

So I wrapped them in the original package and went down to the lower passenger zone, where I found the ship's third officer. I told him that I had some secret diplomatic matter to be destroyed and he took me to the engine room. I shoved the package into one of the mass–energy convertors and watched it resolve itself into its constituent protons, neutrons and electrons.

On the way back, I stopped in at the ship's bar.

Hoddy Ringo was there, wrapped up in—and I use the words literally—a young lady from the Alderbaran system. She was on her way home from one of the quickie divorce courts on Terra and was celebrating her marital emancipation. They were so entangled with each other that they didn't notice me. When they left the bar, I slipped after them until I saw them enter the lady's stateroom. That, of course, would have Hoddy immobilized—better word, located—for a while. So I went back to our suite, picked the lock of Hoddy's room, and allowed myself half an hour to search his luggage.

All of his clothes were new, but there were not a great many of them. Evidently he was planning to re–outfit himself on New Texas. There were a few odds and ends, the kind any man with a real home planet will hold on to, in the luggage.

He had another eleven–mm pistol, made by Consolidated–Martian Metalworks, mate to the one he was carrying in a shoulder–holster, and a wide two–holster belt like the one furnished me, but quite old.

I greeted the sight and the meaning of the old holsters with joy: they weren't the State Department Special Services type. That meant that Hoddy was just one of Natalenko's run–of–the–gallows cutthroats, not important enough to be issued the secret equipment.

But I was a little worried over what I found hidden in the lining of one of his bags, a letter addressed to Space–Commander Lucius C. Stonehenge,

Aggression Department Attaché, New Austin Embassy. I didn't have either the time or the equipment to open it. But, knowing our various Departments, I tried to reassure myself with the thought that it was only a letter–of–credence, with the real message to be delivered orally.

About the real message I had no doubts: *arrange the murder of Ambassador Stephen Silk in such a way that it looks like another New Texan job....*

Starting that evening—or what passed for evening aboard a ship in hyperspace—Hoddy and I began a positively epochal binge together.

I had it figured this way: as long as we were on board ship, I was perfectly safe. On the ship, in fact, Hoddy would definitely have given his life to save mine. I'd have to be killed on New Texas to give Kling's boys their excuse for moving in.

And there was always the chance, with no chance too slender for me to ignore, that I might be able to get Hoddy drunk enough to talk, yet still be sober enough myself to remember what he said.

Exact times, details, faces, names, came to me through a sort of hazy blur as Hoddy and I drank something he called superbourbon—a New Texan drink that Bourbon County, Kentucky, would never have recognized. They had no corn on New Texas. This stuff was made out of something called superyams.

There were at least two things I got out of the binge. First, I learned to slug down the national drink without batting an eye. Second, I learned to control my expression as I uncovered the fact that everything on New Texas was supersomething.

I was also cautious enough, before we really got started, to leave my belt and guns with the purser. I didn't want Hoddy poking around those secret holsters. And I remember telling the captain to radio New Austin as soon as we came out of our last hyperspace–jump, then to send the ship's doctor around to give me my hangover treatments.

But the one thing I wanted to remember, as the hangover shots brought me back to normal life, I found was the one thing I couldn't remember. What was the name of that girl—a big, beautiful blond—who joined the party along with Hoddy's grass widow from Alderbaran and stayed with it to the end?

Damn, I wished I could remember her name!

When we were fifteen thousand miles off–planet and the lighters from New Austin spaceport were reported on the way, I got into the skin–tight Levis, the cataclysmic–colored shirt, and the loose vest, tucked my big hat under my arm, and went to the purser's office for my guns, buckling them on. When I got back to the suite, Hoddy had put on his pistols and was practicing quick draws in front of the mirror. He took one look at my armament and groaned.

"You're gonna get yourself killed for sure, with that rig, an' them popguns," he told me.

"These popguns'll shoot harder and make bigger holes than that pair of museum–pieces you're carrying," I replied.

"An' them holsters!" Hoddy continued. "Why, it'd take all day to get your guns outa them! You better let me find you a real rig, when we get to New Austin...."

There was a chance, of course, that he knew what I was using and wanted to hide his knowledge. I doubted that.

"Sure, you State Department guys always know everything," he went on. "Like them microfilm–books you was readin'. I try to tell you what things is really like on New Texas, an' you let it go in one ear an' out the other."

Then he wandered off to say good–bye to the grass widow from Alderbaran, leaving me to make the last–minute check on the luggage. I was hoping I'd be able to see that blond ... what *was* her name; Gail something–or–other. Let's see, she'd been at some Terran university, and she was on her way home to ... to New Texas! Of course!

I saw her, half an hour later, in the crowd around the airlock when the lighters came alongside, and I tried to push my way toward her. As I did, the airlock opened, the crowd surged toward it, and she was carried along. Then the airlock closed, after she had passed through and before I could get to it. That meant I'd have to wait for the second lighter.

So I made the best of it, and spent the next half–hour watching the disc of the planet grow into a huge ball that filled the lower half of the viewscreen and then lose its curvature, and instead of moving in toward the planet, we were going down toward it.

Chapter III

New Austin spaceport was a huge place, a good fifty miles outside the city. As we descended, I could see that it was laid out like a wheel, with the landings and the blast–off stands around the hub, and high buildings—packing houses and refrigeration plants—along the many spokes. It showed a technological level quite out of keeping with the accounts I had read, or the stories Hoddy had told, about the simple ranch life of the planet. Might be foreign capital invested there, and I made a mental note to find out whose.

On the other hand, Old Texas, on Terra, had been heavily industrialized; so much so that the state itself could handle the gigantic project of building enough spaceships to move almost the whole population into space.

Then the landing–field was rushing up at us, with the nearer ends of the roadways and streets drawing close and the far ends lengthening out away from us. The other lighter was already down, and I could see a crowd around it.

There was a crowd waiting for us when we got out and went down the escalators to the ground, and as I had expected, a special group of men waiting for me. They were headed by a tall, slender individual in the short black Eisenhower jacket, gray–striped trousers and black homburg that was the uniform of the Diplomatic Service, alias the Cookie Pushers.

Over their heads at the other rocket–boat, I could see the gold-gleaming head of the girl I'd met on the ship.

I tried to push through the crowd and get to her. As I did, the Cookie Pusher got in my way.

"Mr. Silk! Mr. Ambassador! Here we are!" he was clamoring. "The car for the Embassy is right over here!" He clutched my elbow. "You have no idea how glad we all are to see you, Mr. Ambassador!"

"Yes, yes; of course. Now, there's somebody over there I have to see, at once." I tried to pull myself loose from his grasp.

Across the concrete between the two lighters, I could see the girl push out of the crowd around her and wave a hand to me. I tried to yell to her; but just then another lighter, loaded with freight, started to lift out at another nearby stand, with the roar of half a dozen Niagaras. The thin man in the striped trousers added to the uproar by shouting into my ear and pulling at me.

"We haven't time!" he finally managed to make himself heard. "We're dreadfully late now, sir! You must come with us."

Hoddy, too, had caught hold of me by the other arm.

"Come on, boss. There's gotta be some reason why he's got himself in an uproar about whatever it is. You'll see her again."

Then, the whole gang—Hoddy, the thin man with the black homburg, his younger accomplice in identical garb, and the chauffeur—all closed in on me and pushed me, pulled me, half–carried me, fifty yards across the concrete to where their air–car was parked. By this time, the tall blond had gotten clear of the mob around her and was waving frantically at me. I tried to wave back, but I was literally crammed into the car and flung down on the seat. At the same time, the chauffeur was jumping in, extending the car's wings, jetting up.

"Great God!" I bellowed. "This is the damnedest piece of impudence I've ever had to suffer from any subordinates in my whole State Department experience! I want an explanation out of you, and it'd better be a good one!"

There was a deafening silence in the car for a moment. The thin man moved himself off my lap, then sat there looking at me with the heartbroken eyes of a friendly dog that had just been kicked for something which wasn't really its fault.

"Mr. Ambassador, you can't imagine how sorry we all are, but if we hadn't gotten you away from the spaceport and to the Embassy at once, we would all have been much sorrier."

"Somebody here gunnin' for the Ambassador?" Hoddy demanded sharply.

"Oh, no! I hadn't even thought of that," the thin man almost gibbered. "But your presence at the Embassy is of immediate and urgent necessity. You have no idea of the state into which things have gotten.... Oh, pardon me, Mr. Ambassador. I am Gilbert W. Thrombley, your chargé d'affaires." I shook hands with him. "And Mr. Benito Gomez, the Secretary of the Embassy." I shook hands with him, too, and started to introduce Mr. Hoddy Ringo.

Hoddy, however, had turned to look out the rear window; immediately, he gave a yelp.

"We got a tail, boss! Two of them! Look back there!"

There were two black eight–passenger aircars, of the same model, whizzing after us, making an obvious effort to overtake us. The chauffeur cursed and fired his auxiliary jets, then his rocket–booster.

Immediately, black rocket-fuel puffs shot away from the pursuing aircars.

Hoddy turned in his seat, cranked open a porthole–slit in the window, and poked one of his eleven–mm's out, letting the whole clip go. Thrombley and Gomez slid down onto the floor, and both began trying to drag me down with them, imploring me not to expose myself.

As far as I could see, there was nothing to expose myself to. The other cars kept coming, but neither of them were firing at us. There was also no indication that Hoddy's salvo had had any effect on them. Our chauffeur went into a perfect frenzy of twisting and dodging, at the same time using his radiophone to tell somebody to get the goddamn gate open in a hurry. I saw the blue skies and green plains of New Texas replacing one another above, under, in front of and behind us. Then the car set down on a broad stretch of concrete, the wings were retracted, and we went whizzing down a city street.

We whizzed down a number of streets. We cut corners on two wheels, and on one wheel, and, I was prepared to swear, on no wheels. A couple of times, with the wings retracted, we actually jetted into the air and jumped over vehicles in front of us, landing again with bone–shaking jolts. Then we made an abrupt turn and shot in under a concrete arch, and a big door banged shut behind us, and we stopped, in the middle of a wide patio, the front of the car a few inches short of a fountain. Four or five people, in diplomatic striped trousers, local dress and the uniform of the Space Marines, came running over.

Thrombley pulled himself erect and half–climbed, half–fell, out of the car. Gomez got out on the other side with Hoddy; I climbed out after Thrombley.

A tall, sandy–haired man in the uniform of the Space Navy came over.

"What the devil's the matter, Thrombley?" he demanded. Then, seeing me, he gave me as much of a salute as a naval officer will ever bestow on anybody in civilian clothes.

"Mr. Silk?" He looked at my costume and the pistols on my belt in well–bred concealment of surprise. "I'm your military attaché, Stonehenge; Space–Commander, Space Navy."

I noticed that Hoddy's ears had pricked up, but he wasn't making any effort to attract Stonehenge's attention. I shook hands with him, introduced Hoddy, and offered my cigarette case around.

"You seem to have had a hectic trip from the spaceport, Mr. Ambassador. What happened?"

Thrombley began accusing our driver of trying to murder the lot of us. Hoddy brushed him aside and explained:

"Just after we'd took off, two other cars took off after us. We speeded up, and they speeded up, too. Then your fly–boy, here, got fancy. That shook 'em off. Time we got into the city, we'd dropped them. Nice job of driving. Probably saved our lives."

"Shucks, that wasn't nothin'," the driver disclaimed. "When you drive for politicians, you're either good or you're good and dead."

"I'm surprised they started so soon," Stonehenge said. Then he looked around at my fellow–passengers, who seemed to have realized, by now, that they were no longer dangling by their fingernails over the brink of the grave. "But gentlemen, let's not keep the Ambassador standing out here in the hot sun."

So we went over the arches at the side of the patio, and were about to sit down when one of the Embassy servants came up, followed by a man in a loose vest and blue Levis and a big hat. He had a pair of automatics in his belt, too.

"I'm Captain Nelson; New Texas Rangers," he introduced himself. "Which one of you–all is Mr. Stephen Silk?"

I admitted it.

The Ranger pushed back his wide hat and grinned at me.

"I just can't figure this out," he said. "You're in the right place and the right company, but we got a report, from a mighty good source, that you'd been kidnapped at the spaceport by a gang of thugs!"

"A blond source?" I made curving motions with my hands. "I don't blame her. My efficient and conscientious chargé d'affaires, Mr. Thrombley, felt that I should reach the Embassy, here, as soon as possible, and from where she was standing, it must have looked like a kidnapping. Fact is, it looked like one from where I was standing, too. Was that you and your people who were chasing us? Then I must apologize for opening fire on you ... I hope nobody was hurt."

"No, our cars are pretty well armored. You scored a couple of times on one of them, but no harm done. I reckon after what happened to Silas Cumshaw, you had a right to be suspicious."

I noticed that refreshments, including several bottles, had been placed on a big wicker table under the arched veranda.

"Can I offer you a drink, Captain, in token of mutual amity?" I asked.

"Well, now, I'd like to, Mr. Ambassador, but I'm on duty ... " he began.

"You can't be. You're an officer of the Planetary Government of New Texas, and in this Embassy, you're in the territory of the Solar League."

"That's right, now, Mr. Ambassador," he grinned. "Extraterritoriality. Wonderful thing, extraterritoriality." He looked at Hoddy, who, for the first time since I had met him, was trying to shrink into the background.

"And diplomatic immunity, too. Ain't it, Hoddy?"

After he had had his drink and departed, we all sat down. Thrombley began speaking almost at once.

"Mr. Ambassador, you must, you simply must, issue a public statement, immediately, sir. Only a public statement, issued promptly, will relieve the crisis into which we have all been thrust."

"Oh, come, Mr. Thrombley," I objected. "Captain Nelson'll take care of all that in his report to his superiors."

Thrombley looked at me for a moment as though I had been speaking to him in Hottentot, then waved his hands in polite exasperation.

"Oh, no, no! I don't mean that, sir. I mean a public statement to the effect that you have assumed full responsibility for the Embassy. Where is that thing? Mr. Gomez!"

Gomez gave him four or five sheets, stapled together. He laid them on the table, turned to the last sheet, and whipped out a pen.

"Here, sir; just sign here."

"Are you crazy?" I demanded. "I'll be damned if I'll sign that. Not till I've taken an inventory of the physical property of the Embassy, and familiarized myself with all its commitments, and had the books audited by some firm of certified public accountants."

Thrombley and Gomez looked at one another. They both groaned.

"But we must have a statement of assumption of responsibility ... " Gomez dithered.

"...or the business of the Embassy will be at a dead stop, and we can't do anything," Thrombley finished.

"Wait a moment, Thrombley," Stonehenge cut in. "I understand Mr. Silk's attitude. I've taken command of a good many ships and installations, at one time or another, and I've never signed for anything I couldn't see and feel and count. I know men who retired as brigadier generals or vice–admirals, but they retired loaded with debts incurred because as second lieutenants or ensigns they forgot that simple rule."

He turned to me. "Without any disrespect to the chargé d'affaires, Mr. Silk, this Embassy has been pretty badly disorganized since Mr. Cumshaw's death. No one felt authorized, or, to put it more accurately, no one dared, to declare himself acting head of the Embassy—"

"Because that would make him the next target?" I interrupted. "Well, that's what I was sent here for. Mr. Gomez, as Secretary of the Embassy, will you please, at once, prepare a statement for the press and telecast release to the effect that I am now the authorized head of this Embassy, responsible from this hour for all its future policies and all its present commitments insofar as they obligate the government of the Solar League. Get that out at once. Tomorrow, I will present my credentials to the Secretary of State here. Thereafter, Mr. Thrombley, you can rest in the assurance that I'll be the one they'll be shooting at."

"But you can't wait that long, Mr. Ambassador," Thrombley almost wailed. "We must go immediately to the Statehouse. The reception for you is already going on."

I looked at my watch, which had been regulated aboard ship for Capella IV time. It was just 1315.

"What time do they hold diplomatic receptions on this planet, Mr. Thrombley?" I asked.

"Oh, any time at all, sir. This one started about 0900 when the news that the ship was in orbit off–planet got in. It'll be a barbecue, of course, and —"

"Barbecued supercow! Yipeee!" Hoddy yelled. "What I been waitin' for for five years!"

It would be the vilest cruelty not to take him along, I thought. And it would also keep him and Stonehenge apart for a while.

"But we must hurry, Mr. Ambassador," Thrombley was saying. "If you will change, now, to formal dress … "

And he was looking at me, gasping. I think it was the first time he had actually seen what I was wearing.

"In native dress, Mr. Ambassador!"

Thrombley's eyes and tone were again those of an innocent spaniel caught in the middle of a marital argument.

Then his gaze fell to my belt and his eyes became saucers. "Oh, dear! And armed!"

My chargé d'affaires was shuddering and he could not look directly at me.

"Mr. Ambassador, I understand that you were recently appointed from the Consular Service. I sincerely hope that you will not take it amiss if I point out, here in private, that—"

"Mr. Thrombley, I am wearing this costume and these pistols on the direct order of Secretary of State Ghopal Singh."

That set him back on his heels.

"I … I can't believe it!" he exclaimed. "An ambassador is *never* armed."

"Not when he's dealing with a government which respects the comity of nations and the usages of diplomatic practice, no," I replied. "But the fate of Mr. Cumshaw clearly indicates that the government of New Texas is not such a government. These pistols are in the nature of a not–too–subtle hint of the manner in which this government, here, is being regarded by the government of the Solar League." I turned to Stonehenge. "Commander, what sort of an Embassy guard have we?" I asked.

"Space Marines, sergeant and five men. I double as guard officer, sir."

"Very well. Mr. Thrombley insists that it is necessary for me to go to this fish–fry or whatever it is immediately. I want two men, a driver and an auto–rifleman, for my car. And from now on, I would suggest, Commander, that you wear your sidearm at all times outside the Embassy."

"Yes, sir!" and this time, Stonehenge gave me a real salute.

"Well, I must phone the Statehouse, then," Thrombley said. "We will have to call on Secretary of State Palme, and then on President Hutchinson."

With that, he got up, excused himself, motioned Gomez to follow, and hurried away.

I got up, too, and motioned Stonehenge aside.

"Aboard ship, coming in, I was told that there's a task force of the Space Navy on maneuvers about five light–years from here," I said.

"Yes, sir. Task Force Red–Blue–Green, Fifth Space Fleet. Fleet Admiral Sir Rodney Tregaskis."

"Can we get hold of a fast space–boat, with hyperdrive engines, in a hurry?"

"Eight or ten of them always around New Austin spaceport, available for charter."

"All right; charter one and get out to that fleet. Tell Admiral Tregaskis that the Ambassador at New Austin feels in need of protection; possibility of z'Srauff invasion. I'll give you written orders. I want the Fleet within radio call. How far out would that be, with our facilities?"

"The Embassy radio isn't reliable beyond about sixty light–minutes, sir."

"Then tell Sir Rodney to bring his fleet in that close. The invasion, if it comes, will probably not come from the direction of the z'Srauff star–cluster; they'll probably jump past us and move in from the other side. I hope you don't think I'm having nightmares, Commander. Danger of a z'Srauff invasion was pointed out to me by persons on the very highest level, on Luna."

Stonehenge nodded. "I'm always having the same kind of nightmares, sir. Especially since this special envoy arrived here, ostensibly to negotiate a meteor–mining treaty." He hesitated for a moment. "We don't want the New Texans to know, of course, that you've sent for the fleet?"

"Naturally not."

"Well, if I can wait till about midnight before I leave, I can get a boat owned, manned and operated by Solar League people. The boat's a dreadful–looking old tub, but she's sound and fast. The gang who own her are pretty notorious characters—suspected of smuggling, piracy, and what not—but they'll keep their mouths shut if well paid."

"Then pay them well," I said. "And it's just as well you're not leaving at once. When I get back from this clambake, I'll want to have a general informal council, and I certainly want you in on it."

On the way to the Statehouse in the aircar, I kept wondering just how smart I had been.

I was pretty sure that the z'Srauff was getting ready for a sneak attack on New Texas, and, as Solar League Ambassador, I of course had the right to call on the Space Navy for any amount of armed protection.

Sending Stonehenge off on what couldn't be less than an eighteen–hour trip would delay anything he and Hoddy might be cooking up, too.

On the other hand, with the fleet so near, they might decide to have me rubbed out in a hurry, to justify seizing the planet ahead of the z'Srauff.

I was in that pleasant spot called, "Damned if you do and damned if you don't...."

Chapter IV

The Statehouse appeared to cover about a square mile of ground and it was an insane jumble of buildings piled beside and on top of one another, as though it had been in continuous construction ever since the planet was colonized, eighty–odd years before.

At what looked like one of the main entrances, the car stopped. I told our Marine driver and auto–rifleman to park the car and take in the barbecue, but to leave word with the doorman where they could be found. Hoddy, Thrombley and I then went in, to be met by a couple of New Texas Rangers, one of them the officer who had called at the Embassy. They guided us to the office of the Secretary of State.

"We're dreadfully late," Thrombley was fretting. "I do hope we haven't kept the Secretary waiting too long."

From the looks of him, I was afraid we had. He jumped up from his desk and hurried across the room as soon as the receptionist opened the door for us, his hand extended.

"Good afternoon, Mr. Thrombley," he burbled nervously. "And this is the new Ambassador, I suppose. And this—" He caught sight of Hoddy Ringo, bringing up the rear and stopped short, hand flying to open mouth. "Oh, dear me!"

So far, I had been building myself a New Texas stereotype from Hoddy Ringo and the Ranger officer who had chased us to the Embassy. But this frightened little rabbit of a fellow simply didn't fit it. An alien would be justified in assigning him to an entirely different species.

Thrombley introduced me. I introduced Hoddy as my confidential secretary and advisor. We all shook hands, and Thrombley dug my credentials out of his briefcase and handed them to me, and I handed them to the Secretary of State, Mr. William A. Palme. He barely glanced at them, then shook my hand again fervently and mumbled something about "inexpressible pleasure" and "entirely acceptable to my government."

That made me the accredited and accepted Ambassador to New Texas.

Mr. Palme hoped, or said he hoped, that my stay in New Texas would be long and pleasant. He seemed rather less than convinced that it would be. His eyes kept returning in horrified fascination to my belt. Each time they would focus on the butts of my Krupp–Tattas, he would pull them resolutely away again.

"And now, we must take you to President Hutchinson; he is most anxious to meet you, Mr. Silk. If you will please come with me ... "

Four or five Rangers who had been loitering the hall outside moved to follow us as we went toward the elevator. Although we had come into the building onto a floor only a few feet above street–level, we went down three floors from the hallway outside the Secretary of State's office, into a huge room, the concrete floor of which was oil–stained, as though vehicles were continually being driven in and out. It was about a hundred feet wide, and two or three hundred in length. Daylight was visible through open doors at the end. As we approached them, the Rangers fanning out on either side and in front of us, I could hear a perfect bedlam of noise outside —shouting, singing, dance–band music, interspersed with the banging of shots.

When we reached the doors at the end, we emerged into one end of a big rectangular plaza, at least five hundred yards in length. Most of the uproar was centered at the opposite end, where several thousand people, in costumes colored through the whole spectrum, were milling about. There seemed to be at least two square–dances going on, to the music of competing bands. At the distant end of the plaza, over the heads of the crowd, I could see the piles and tracks of an overhead crane, towering above what looked like an open–hearth furnace. Between us and the bulk of the crowd, in a cleared space, two medium tanks, heavily padded with mats, were ramming and trying to overturn each other, the mob of spectators crowding as close to them as they dared. The din was positively deafening, though we were at least two hundred yards from the center of the crowd.

"Oh, dear, I always dread these things!" Palme was saying.

"Yes, absolutely anything could happen," Thrombley twittered.

"Man, this is a real barbecue!" Hoddy gloated. "Now I really feel at home!"

"Over this way, Mr. Silk," Palme said, guiding me toward the short end of the plaza, on our left. "We will see the President and then ... "

He gulped.

"...then we will all go to the barbecue."

In the center of the short end of the plaza, dwarfed by the monster bulks of steel and concrete and glass around it, stood a little old building of warm-tinted adobe. I had never seen it before, but somehow it was familiar-looking. And then I remembered. Although I had never seen it before, I had seen it pictured many times; pictured under attack, with gunsmoke spouting from windows and parapets.

I plucked Thrombley's sleeve.

"Isn't that a replica of the Alamo?"

He was shocked. "Oh, dear, Mr. Ambassador, don't let anybody hear you ask that. That's no replica. It *is* the Alamo. *The* Alamo."

I stood there a moment, looking at it. I was remembering, and finally understanding, what my psycho–history lessons about the "Romantic Freeze" had meant.

They had taken this little mission–fort down, brick by adobe brick, loaded it carefully into a spaceship, brought it here, forty two light–years away from Terra, and reverently set it up again. Then they had built a whole world and a whole social philosophy around it.

It had been the dissatisfied, of course, the discontented, the dreamers, who had led the vanguard of man's explosion into space following the discovery of the hyperspace–drive. They had gone from Terra cherishing dreams of things that had been dumped into the dust bin of history, carrying with them pictures of ways of life that had passed away, or that had never really been. Then, in their new life, on new planets, they had set to work making those dreams and those pictures live.

And, many times, they had come close to succeeding.

These Texans, now: they had left behind the cold fact that it had been their state's great industrial complex that had made their migration possible. They ignored the fact that their life here on Capella IV was possible only by application of modern industrial technology. That rodeo down the plaza —tank–tilting instead of bronco–busting. Here they were living frozen in a romantic dream, a world of roving cowboys and ranch kingdoms.

No wonder Hoddy hadn't liked the books I had been reading on the ship. They shook the fabric of that dream.

There were people moving about, at this relatively quiet end of the plaza, mostly in the direction of the barbecue. Ten or twelve Rangers loitered at the front of the Alamo, and with them I saw the dress blues of my two Marines. There was a little three–wheeled motorcart among them, from which they were helping themselves to food and drink. When they saw us coming, the two Marines shoved their sandwiches into the hands of a couple of Rangers and tried to come to attention.

"At ease, at ease," I told them. "Have a good time, boys. Hoddy, you better get in on some of this grub; I may be inside for quite a while."

As soon as the Rangers saw Hoddy, they hastily got things out of their right hands. Hoddy grinned at them.

"Take it easy, boys," he said. "I'm protected by the game laws. I'm a diplomat, I am."

There were a couple of Rangers lounging outside the door of the President's office and both of them carried autorifles, implying things I didn't like.

I had seen the President of the Solar League wandering around the dome–city of Artemis unattended, looking for all the world like a professor in his academic halls. Since then, maybe before then, I had always had a healthy suspicion of governments whose chiefs had to surround themselves with bodyguards.

But the President of New Texas, John Hutchinson, was alone in his office when we were shown in. He got up and came around his desk to greet us, a slender, stoop–shouldered man in a black–and–gold laced jacket. He had a narrow compressed mouth and eyes that seemed to be watching every corner of the room at once. He wore a pair of small pistols in cross–body holsters under his coat, and he always kept one hand or the other close to his abdomen.

He was like, and yet unlike, the Secretary of State. Both had the look of hunted animals; but where Palme was a rabbit, twitching to take flight at the first whiff of danger, Hutchinson was a cat who hears hounds baying— ready to run if he could, or claw if he must.

"Good day, Mr. Silk," he said, shaking hands with me after the introductions. "I see you're heeled; you're smart. You wouldn't be here today if poor Silas Cumshaw'd been as smart as you are. Great man, though; a wise and farseeing statesman. He and I were real friends."

"You know who Mr. Silk brought with him as bodyguard?" Palme asked. "Hoddy Ringo!"

"Oh, my God! I thought this planet was rid of him!" The President turned to me. "You got a good trigger–man, though, Mr. Ambassador. Good man to watch your back for you. But lot of folks here won't thank you for bringing him back to New Texas."

He looked at his watch. "We have time for a little drink, before we go outside, Mr. Silk," he said. "Care to join me?"

I assented and he got a bottle of superbourbon out of his desk, with four glasses. Palme got some water tumblers and brought the pitcher of ice–water from the cooler.

I noticed that the New Texas Secretary of State filled his three–ounce liquor glass to the top and gulped it down at once. He might act as though he were descended from a long line of maiden aunts, but he took his liquor in blasts that would have floored a spaceport labor–boss.

We had another drink, a little slower, and chatted for a while, and then Hutchinson said, regretfully that we'd have to go outside and meet the folks. Outside, our guards—Hoddy, the two Marines, the Rangers who had escorted us from Palme's office, and Hutchinson's retinue—surrounded us, and we made our way down the plaza, through the crowd. The din—ear–

piercing yells, whistles, cowbells, pistol shots, the cacophony of the two dance-bands, and the chorus-singing, of which I caught only the words: *The skies of freedom are above you!*—was as bad as New Year's Eve in Manhattan or Nairobi or New Moscow, on Terra.

"Don't take all this as a personal tribute, Mr. Silk!" Hutchinson screamed into my ear. "On this planet, to paraphrase Nietzsche, a good barbecue halloweth any cause!"

That surprised me, at the moment. Later I found out that John Hutchinson was one of the leading scholars on New Texas and had once been president of one of their universities. New Texas Christian, I believe.

As we got up onto the platform, close enough to the barbecue pits to feel the heat from them, somebody let off what sounded like a fifty-mm anti-tank gun five or six times. Hutchinson grabbed a microphone and bellowed into it: "Ladies and gentlemen! Your attention, please!"

The noise began to diminish, slowly, until I could hear one voice, in the crowd below:

"Shut up, you damn fools! We can't eat till this is over!"

Hutchinson introduced me, in very few words. I gathered that lengthy speeches at barbecues were not popular on New Texas.

"Ladies and gentlemen! ' I yelled into the microphone. "Appreciative as I am of this honor, there is one here who is more deserving of your notice than I; one to whom I, also, pay homage. He's over there on the fire, and I want a slice of him as soon as possible!"

That got a big ovation. There was, beside the water pitcher, a bottle of superbourbon. I ostentatiously threw the water out of the glass, poured a big shot of the corrosive stuff, and downed it.

"For God's sake, let's eat!" I finished. Then I turned to Thrombley, who was looking like a priest who has just seen the bishop spit in the holy-water font. "Stick close to me," I whispered. "Cue me in on the local notables, and the other members of the Diplomatic Corps." Then we all got down off the platform, and a band climbed up and began playing one of those raucous "cowboy ballads" which had originated in Manhattan about the middle of the Twentieth Century.

"The sandwiches'll be here in a moment, Mr. Ambassador," Hutchinson screamed—in effect, whispered—in my ear. "Don't feel any reluctance about shaking hands with a sandwich in your other hand; that's standard practice, here. You struck just the right note, up there. That business with the liquor was positively inspired!"

The sandwiches—huge masses of meat and hot relish, wrapped in tortillas of some sort—arrived and I bit into one.

I'd been eating supercow all my life, frozen or electron–beamed for transportation, and now I was discovering that I had never really eaten supercow before. I finished the first sandwich in surprisingly short order and was starting on my second when the crowd began coming.

First, the Diplomatic Corps, the usual collection of weirdies, human and otherwise....

There was the Ambassador from Tara, in a suit of what his planet produced as a substitute for Irish homespuns. His Embassy, if it was like the others I had seen elsewhere, would be an outsize cottage with whitewashed walls and a thatched roof, with a bowl of milk outside the door for the Little People ...

The Ambassador from Alpheratz II, the South African Nationalist planet, with a full beard, and old fashioned plug hat and tail–coat. They were a frustrated lot. They had gone into space to practice *apartheid* and had settled on a planet where there was no other intelligent race to be superior to....

The Mormon Ambassador from Deseret—Delta Camelopardalis V....

The Ambassador from Spica VII, a short jolly–looking little fellow, with a head like a seal's, long arms, short legs and a tail like a kangaroo's....

The Ambassador from Beta Cephus VI, who could have passed for human if he hadn't had blood with a copper base instead of iron. His skin was a dark green and his hair was a bright blue....

I was beginning to correct my first impression that Thrombley was a complete dithering fool. He stood at my left elbow, whispering the names and governments and home planets of the Ambassadors as they came up, handing me little slips of paper on which he had written phonetically correct renditions of the greetings I would give them in their own language. I was still twittering a reply to the greeting of Nanadabadian, from Beta Cephus VI, when he whispered to me:

"Here it comes, sir. The z'Srauff!"

The z'Srauff were reasonably close to human stature and appearance, allowing for the fact that their ancestry had been canine instead of simian. They had, of course, longer and narrower jaws than we have, and definitely carnivorous teeth.

There were stories floating around that they enjoyed barbecued Terran even better than they did supercow and hot relish.

This one advanced, extending his three-fingered hand.

"I am most happy to make connection with Solar League representative," he said. "I am named Gglafrr Ddespttann Vuvuvu."

No wonder Thrombley let him introduce himself. I answered in the Basic English that was all he'd admit to understanding:

"The name of your great nation has gone before you to me. The stories we tell to our young of you are at the top of our books. I have hope to make great pleasure in you and me to be friends."

Gglafrr Vuvuvu's smile wavered a little at the oblique reference to the couple of trouncings our Space Navy had administered to z'Srauff ships in the past. "We will be in the same place again times with no number," the alien replied. "I have hope for you that time you are in this place will be long and will put pleasure in your heart."

Then the pressure of the line behind him pushed him on. Cabinet Members; Senators and Representatives; prominent citizens, mostly Judge so–and–so, or Colonel this–or–that. It was all a blur, so much so that it was an instant before I recognized the gleaming golden hair and the statuesque figure.

"Thank you! I have met the Ambassador." The lovely voice was shaking with restrained anger.

"Gail!" I exclaimed.

"Your father coming to the barbecue, Gail?" President Hutchinson was asking.

"He ought to be here any minute. He sent me on ahead from the hotel. He wants to meet the Ambassador. That's why I joined the line."

"Well, suppose I leave Mr. Silk in your hands for a while," Hutchinson said. "I ought to circulate around a little."

"Yes. Just leave him in my hands!" she said vindictively.

"What's wrong, Gail?" I wanted to know. "I know, I was supposed to meet you at the spaceport, but—"

"You made a beautiful fool of me at the spaceport!"

"Look, I can explain everything. My Embassy staff insisted on hurrying me off—"

Somebody gave a high–pitched whoop directly behind me and emptied the clip of a pistol. I couldn't even hear what else I said. I couldn't hear what she said, either, but it was something angry.

"You have to listen to me!" I roared in her ear. "I can explain everything!"

"Any diplomat can explain anything!" she shouted back.

"Look, Gail, you're hanging an innocent man!" I yelled back at her. "I'm entitled to a fair trial!"

Somebody on the platform began firing his pistol within inches of the loud–speakers and it sounded like an H–bomb going off. She grabbed my wrist and dragged me toward a door under the platform.

"Down here!" she yelled. "And this better be good, Mr. Silk!"

We went down a spiral ramp, lighted by widely–scattered overhead lights.

"Space–attack shelter," she explained. "And look: what goes on in space–ships is one thing, but it's as much as a girl's reputation is worth to come down here during a barbecue."

There seemed to be quite few girls at that barbecue who didn't care what happened to their reputations. We discovered that after looking into a couple of passageways that branched off the entrance.

"Over this way," Gail said, "Confederate Courts Building. There won't be anything going on over here, now."

I told her, with as much humorous detail as possible, about how Thrombley had shanghaied me to the Embassy, and about the chase by the Rangers. Before I was half through, she was laughing heartily, all traces of her anger gone. Finally, we came to a stairway, and at the head of it to a small door.

"It's been four years that I've been away from here," she said. "I think there's a reading room of the Law Library up here. Let's go in and enjoy the quiet for a while."

But when we opened the door, there was a Ranger standing inside.

"Come to see a trial, Mr. Silk? Oh, hello, Gail. Just in time; they're going to prepare for the next trial."

As he spoke, something clicked at the door. Gail looked at me in consternation.

"Now we're locked in," she said. "We can't get out till the trial's over."

Chapter V

I looked around.

We were on a high balcony, at the end of a long, narrow room. In front of us, windows rose to the ceiling, and it was evident that the floor of the room was about twenty feet below ground level. Outside, I could see the barbecue still going on, but not a murmur of noise penetrated to us. What seemed to be the judge's bench was against the outside wall, under the tall windows. To the right of it was a railed stand with a chair in it, and in front, arranged in U–shape, were three tables at which a number of men were hastily conferring. There were nine judges in a row on the bench, all in black gowns. The spectators' seats below were filled with people, and there were quite a few up here on the balcony.

"What is this? Supreme Court?" I asked as Gail piloted me to a couple of seats where we could be alone.

"No, Court of Political Justice," she told me. "This is the court that's going to try those three Bonney brothers, who killed Mr. Cumshaw."

It suddenly occurred to me that this was the first time I had heard anything specific about the death of my predecessor.

"That isn't the trial that's going on now, I hope?"

"Oh, no; that won't be for a couple of days. Not till after you can arrange to attend. I don't know what this trial is. I only got home today, myself."

"What's the procedure here?" I wanted to know.

"Well, those nine men are judges," she began. "The one in the middle is President Judge Nelson. You've met his son—the Ranger officer who chased you from the spaceport. He's a regular jurist. The other eight are prominent citizens who are drawn from a panel, like a jury. The men at the table on the left are the prosecution: friends of the politician who was killed. And the ones on the right are the defense: they'll try to prove that the dead man got what was coming to him. The ones in the middle are friends of the court: they're just anybody who has any interest in the case —people who want to get some point of law cleared up, or see some precedent established, or something like that."

"You seem to assume that this is a homicide case," I mentioned.

"They generally are. Sometimes mayhem, or wounding, or simple assault, but—"

There had been some sort of conference going on in the open space of floor between the judges' bench and the three tables. It broke up, now, and the judge in the middle rapped with his gavel.

"Are you gentlemen ready?" he asked. "All right, then. Court of Political Justice of the Confederate Continents of New Texas is now in session. Case of the friends of S. Austin Maverick, deceased, late of James Bowie Continent, versus Wilbur Whately."

"My God, did somebody finally kill Aus Maverick?" Gail whispered.

On the center table, in front of the friends of the court, both sides seemed to have piled their exhibits; among the litter I saw some torn clothing, a big white sombrero covered with blood, and a long machete.

"The general nature of the case," the judge was saying, "is that the defendant, Wilbur Whately, of Sam Houston Continent, is here charged with divers offenses arising from the death of the Honorable S. Austin Maverick, whom he killed on the front steps of the Legislative Assembly Building, here in New Austin...."

What goes on here? I thought angrily. *This is the rankest instance of a pre–judged case I've ever seen.* I started to say as much to Gail, but she hushed me.

"I want to hear the specifications," she said.

A man at the prosecution table had risen.

"Please the court," he began, "the defendant, Wilbur Whately, is here charged with political irresponsibility and excessive atrocity in exercising his constitutional right of criticism of a practicing politician.

"The specifications are, as follows: That, on the afternoon of May Seventh, Anno Domini 2193, the defendant here present did arm himself with a machete, said machete not being one of his normal and accustomed weapons, and did loiter in wait on the front steps of the Legislative Assembly Building in the city of New Austin, Continent of Sam Houston, and did approach the decedent, addressing him in abusive, obscene, and indecent language, and did set upon and attack him with the machete aforesaid, causing the said decedent, S. Austin Maverick, to die."

The court wanted to know how the defendant would plead. Somebody, without bothering to rise, said, "Not guilty, Your Honor," from the defense table.

There was a brief scraping of chairs; four of five men from the defense and the prosecution tables got up and advanced to confer in front of the bench, comparing sheets of paper. The man who had read the charges, obviously the chief prosecutor, made himself the spokesman.

"Your Honor, defense and prosecution wish to enter the following stipulations: That the decedent was a practicing politician within the meaning of the Constitution, that he met his death in the manner stated in the coroner's report, and that he was killed by the defendant, Wilbur Whately."

"Is that agreeable to you, Mr. Vincent?" the judge wanted to know.

The defense answered affirmatively. I sat back, gaping like a fool. Why, that was practically—no, it *was*—a confession.

"All right, gentlemen," the judge said. "Now we have all that out of the way, let's get on with the case."

As though there were any case to get on with! I fully expected them to take it on from there in song, words by Gilbert and music by Sullivan.

"Well, Your Honor, we have a number of character witnesses," the prosecution—prosecution, for God's sake!—announced.

"Skip them," the defense said. "We stipulate."

"But you can't stipulate character testimony," the prosecution argued. "You don't know what our witnesses are going to testify to."

"Sure we do: they're going to give us a big long shaggy-dog story about the Life and Miracles of Saint Austin Maverick. We'll agree in advance to all that; this case is concerned only with his record as a politician. And as he spent the last fifteen years in the Senate, that's all a matter of public record. I assume that the prosecution is going to introduce all that, too?"

"Well, naturally ... " the prosecutor began.

"Including his public acts on the last day of his life?" the counsel for the defense demanded. "His actions on the morning of May seventh as chairman of the Finance and Revenue Committee? You going to introduce that as evidence for the prosecution?"

"Well, now ... " the prosecutor began.

"Your Honor, we ask to have a certified copy of the proceedings of the Senate Finance and Revenue Committee for the morning of May Seventh, 2193, read into the record of this court," the counsel for the defense said. "And thereafter, we rest our case."

"Has the prosecution anything to say before we close the court?" Judge Nelson inquired.

"Well, Your Honor, this seems ... that is, we ought to hear both sides of it. My old friend, Aus Maverick, was really a fine man; he did a lot of good for the people of his continent...."

"Yeah, we'd of lynched him, when he got back, if somebody hadn't chopped him up here in New Austin!" a voice from the rear of the courtroom broke in.

The prosecution hemmed and hawed for a moment, and then announced, in a hasty mumble, that it rested.

"I will now close the court," Judge Nelson said. "I advise everybody to keep your seats. I don't think it's going to be closed very long."

And then, he actually closed the court; pressing a button on the bench, he raised a high black screen in front of him and his colleagues. It stayed up for some sixty seconds, and then dropped again.

"The Court of Political Justice has reached a verdict," he announced. "Wilbur Whately, and your attorney, approach and hear the verdict."

The defense lawyer motioned a young man who had been sitting beside him to rise. In the silence that had fallen, I could hear the defendant's boots squeaking as he went forward to hear his fate. The judge picked up a belt and a pair of pistols that had been lying in front of him.

"Wilbur Whately," he began, "this court is proud to announce that you have been unanimously acquitted of the charge of political irresponsibility, and of unjustified and excessive atrocity.

"There was one dissenting vote on acquitting you of the charge of political irresponsibility; one of the associate judges felt that the late unmitigated scoundrel, Austin Maverick, ought to have been skinned alive, an inch at a time. You are, however, acquitted of that charge, too.

"You all know," he continued, addressing the entire assemblage, "the reason for which this young hero cut down that monster of political iniquity, S. Austin Maverick. On the very morning of his justly–merited death, Austin Maverick, using the powers of his political influence, rammed through the Finance and Revenue Committee a bill entitled 'An Act for the Taxing of Personal Incomes, and for the Levying of a Withholding Tax.' Fellow citizens, words fail me to express my horror of this diabolic proposition, this proposed instrument of tyrannical extortion, borrowed from the Dark Ages of the Twentieth Century! Why, if this young nobleman had not taken his blade in hand, I'd have killed the sonofabitch, myself!"

He leaned forward, extending the belt and holsters to the defendant.

"I therefore restore to you your weapons, taken from you when, in compliance with the law, you were formally arrested. Buckle them on, and, assuming your weapons again, go forth from this court a free man, Wilbur Whately. And take with you that machete with which you vindicated the liberties and rights of all New Texans. Bear it reverently to your home, hang it among your lares and penates, cherish it, and dying, mention it within your will, bequeathing it as a rich legacy unto your issue!

Court adjourned; next session 0900 tomorrow. For Chrissake, let's get out of here before the barbecue's over!"

Some of the spectators, drooling for barbecued supercow, began crowding and jostling toward the exits; more of them were pushing to the front of the courtroom, cheering and waving their hip–flasks. The prosecution and about half of the friends of the court hastily left by a side door, probably to issue statements disassociating themselves from the deceased Maverick.

"So that's the court that's going to try the men who killed Ambassador Cumshaw," I commented, as Gail and I went out. "Why, the purpose of that court seems to be to acquit murderers."

"Murderers?" She was indignant. "That wasn't murder. He just killed a politician. All the court could do was determine whether or not the politician needed it, and while I never heard about Maverick's income–tax proposition, I can't see how they could have brought in any other kind of a verdict. Of all the outrageous things!"

I was thoughtfully silent as we went out into the plaza, which was still a riot of noise and polychromatic costumes. And my thoughts were as weltered as the scene before me.

Apparently, on New Texas, killing a politician wasn't regarded as *mallum in se*, and was *mallum prohibitorum* only to the extent that what happened to the politician was in excess of what he deserved. I began to understand why Palme was such a scared rabbit, why Hutchinson had that hunted look and kept his hands always within inches of his pistols.

I began to feel more pity than contempt for Thrombley, too. *He's been on this planet too long and he should never have been sent here in the first place. I'll rotate him home as soon as possible....*

Then the full meaning of what I had seen finally got through to me: if they were going to try the killers of Cumshaw in that court, that meant that on New Texas, foreign diplomats were regarded as practicing politicians....

That made me a practicing politician too!

And that's why, when we got back to the vicinity of the bandstand, I had my right hand close to my pistol, with my thumb on the inconspicuous little spot of silver inlay that operated the secret holster mechanism.

I saw Hutchinson and Palme and Thrombley ahead. With them was a newcomer, a portly, ruddy–faced gentleman with a white mustache and goatee, dressed in a white suit. Gail broke away from me and ran toward him. This, I thought, would be her father; now I would be introduced and find out just what her last name was. I followed, more slowly, and saw a waiter, with a wheeled serving–table, move in behind the group which she had joined.

So I saw what none of them did—the waiter suddenly reversed his long carving–knife and poised himself for a blow at President Hutchinson's back. I simply pressed the little silver stud on my belt, the Krupp–Tatta popped obediently out of the holster into my open hand. I thumbed off the safety and swung up; when my sights closed on the rising hand that held the knife, I fired.

Hoddy Ringo, who had been holding a sandwich with one hand and a drink with the other, dropped both and jumped on the man whose hand I had smashed. A couple of Rangers closed in and grabbed him, also. The group around President Hutchinson had all turned and were staring from me to the man I had shot, and from him to the knife with the broken handle, lying on the ground.

Hutchinson spoke first. "Well, Mr. Ambassador! My Government thanks your Government! That was nice shooting!"

"Hey, you been holdin' out on me!" Hoddy accused. "I never knew you was that kinda gunfighter!"

"There's a new wrinkle," the man with the white goatee said. "We'll have to screen the help at these affairs a little more closely." He turned to me. "Mr. Ambassador, New Texas owes you a great deal for saving the President's life. If you'll get that pistol out of your hand, I'd be proud to shake it, sir."

I holstered my automatic, and took his hand. Gail was saying, "Stephen, this is my father," and at the same time, Palme, the Secretary of State, was doing it more formally:

"Ambassador Silk, may I present one of our leading citizens and large ranchers, Colonel Andrew Jackson Hickock."

Dumbarton Oaks had taught me how to maintain the proper diplomat's unchanging expression; drinking superbourbon had been a post–graduate course. I needed that training as I finally learned Gail's last name.

Chapter VI

It was early evening before we finally managed to get away from the barbecue. Thrombley had called the Embassy and told them not to wait dinner for us, so the staff had finished eating and were relaxing in the patio when our car came in through the street gate. Stonehenge and another man came over to meet us as we got out—a man I hadn't met before.

He was a little fellow, half–Latin, half–Oriental; in New Texas costume and wearing a pair of pistols like mine, in State Department Special Services holsters. He didn't look like a Dumbarton Oaks product: I thought he was more likely an alumnus of some private detective agency.

"Mr. Francisco Parros, our Intelligence man," Stonehenge introduced him.

"Sorry I wasn't here when you arrived, Mr. Silk," Parros said. "Out checking on some things. But I saw that bit of shooting, on the telecast screen in a bar over town. You know, there was a camera right over the bandstand that caught the whole thing—you and Miss Hickock coming toward the President and his party, Miss Hickock running forward to her father, the waiter going up behind Hutchinson with the knife, and then that beautiful draw and snap shot. They ran it again a couple of times on the half–hourly newscast. Everybody in New Austin, maybe on New Texas, is talking about it, now."

"Yes, indeed, sir," Gomez, the Embassy Secretary, said, joining us. "You've made yourself more popular in the eight hours since you landed than poor Mr. Cumshaw had been able to do in the ten years he spent here. But, I'm afraid, sir, you've given me a good deal of work, answering your fan–mail."

We went over and sat down at one of the big tables under the arches at the side of the patio.

"Well, that's all to the good," I said. "I'm going to need a lot of local good will, in the next few weeks. No thanks, Mr. Parros," I added, as the Intelligence man picked up a bottle and made to pour for me. "I've been practically swimming in superbourbon all afternoon. A little black coffee, if you don't mind. And now, gentlemen, if you'll all be seated, we'll see what has to be done."

"A council of war, in effect, Mr. Ambassador?" Stonehenge inquired.

"Let's call it a council to estimate the situation. But I'll have to find out from you first exactly what the situation here is."

Thrombley stirred uneasily. "But sir, I confess that I don't understand. Your briefing on Luna…."

"Was practically nonexistent. I had a total of six hours to get aboard ship, from the moment I was notified that I had been appointed to this Embassy."

"Incredible!" Thrombley murmured.

I wondered what he'd say if I told him that I thought it was deliberate.

"Naturally, I spent some time on the ship reading up on this planet, but I know practically nothing about what's been going on here in, say, the last year. And all I know about the death of Mr. Cumshaw is that he is said to have been killed by three brothers named Bonney."

"So you'll want just about everything, Mr. Silk," Thrombley said. "Really, I don't know where to begin."

"Start with why and how Mr. Cumshaw was killed. The rest, I believe, will key into that."

So they began; Thrombley, Stonehenge and Parros doing the talking. It came to this:

Ever since we had first established an Embassy on New Texas, the goal of our diplomacy on this planet had been to secure it into the Solar League. And it was a goal which seemed very little closer to realization now than it had been twenty–three years before.

"You must know, by now, what politics on this planet are like, Mr. Silk," Thrombley said.

"I have an idea. One Ambassador gone native, another gone crazy, the third killed himself, the fourth murdered."

"Yes, indeed. I've been here fifteen years, myself...."

"That's entirely too long for anybody to be stationed in this place," I told him. "If I'm not murdered, myself, in the next couple of weeks, I'm going to see that you and any other member of this staff who's been here over ten years are rotated home for a tour of duty at Department Headquarters."

"Oh, would you, Mr. Silk? I would be so happy...."

Thrombley wasn't much in the way of an ally, but at least he had a sound, selfish motive for helping me stay alive. I assured him I would get him sent back to Luna, and then went on with the discussion.

Up until six months ago, Silas Cumshaw had modeled himself after the typical New Texas politician. He had always worn at least two faces, and had always managed to place himself on every side of every issue at once. Nothing he ever said could possibly be construed as controversial. Naturally, the cause of New Texan annexation to the Solar League had made no progress whatever.

Then, one evening, at a banquet, he had executed a complete 180–degree turn, delivering a speech in which he proclaimed that union with the Solar League was the only possible way in which New Texans could retain even a vestige of local sovereignty. He had talked about an invasion as though the enemy's ships were already coming out of hyperspace, and had named the invader, calling the z'Srauff "our common enemy." The z'Srauff Ambassador, also present, had immediately gotten up and stalked out, amid a derisive chorus of barking and baying from the New Texans. The New Texans were first shocked and then wildly delighted; they had been so used to hearing nothing but inanities and high–order abstractions from their public figures that the Solar League Ambassador had become a hero overnight.

"Sounds as though there is a really strong sentiment at what used to be called the grass–roots level in favor of annexation," I commented.

"There is," Parros told me. "Of course, there is a very strong isolationist, anti–annexation, sentiment, too. The sentiment in favor of annexation is based on the point Mr. Cumshaw made—the danger of conquest by the z'Srauff. Against that, of course, there is fear of higher taxes, fear of loss of local sovereignty, fear of abrogation of local customs and institutions, and chauvinistic pride."

"We can deal with some of that by furnishing guarantees of local self-government; the emotional objections can be met by convincing them that we need the great planet of New Texas to add glory and luster to the Solar League," I said. "You think, then, that Mr. Cumshaw was assassinated by opponents of annexation?"

"Of course, sir," Thrombley replied. "These Bonneys were only hirelings. Here's what happened, on the day of the murder:

"It was the day after a holiday, a big one here on New Texas, celebrating some military victory by the Texans on Terra, a battle called San Jacinto. We didn't have any business to handle, because all the local officials were home nursing hangovers, so when Colonel Hickock called—"

"Who?" I asked sharply.

"Colonel Hickock. The father of the young lady you were so attentive to at the barbecue. He and Mr. Cumshaw had become great friends, beginning shortly before the speech the Ambassador made at that banquet. He called about 0900, inviting Mr. Cumshaw out to his ranch for the day, and as there was nothing in the way of official business, Mr. Cumshaw said he'd be out by 1030.

"When he got there, there was an aircar circling about, near the ranchhouse. As Mr. Cumshaw got out of his car and started up the front steps, somebody in this car landed it on the driveway and began shooting with a twenty–mm auto–rifle. Mr. Cumshaw was hit several times, and killed instantly."

"The fellows who did the shooting were damned lucky," Stonehenge took over. "Hickock's a big rancher. I don't know how much you know about supercow–ranching, sir, but those things have to be herded with tanks and light aircraft, so that every rancher has at his disposal a fairly good small air–armor combat team. Naturally, all the big ranchers are colonels in the Armed Reserve. Hickock has about fifteen fast fighters, and thirty medium tanks armed with fifty–mm guns. He also has some AA–guns around his ranch house—every once in a while, these ranchers get to squabbling among themselves.

"Well, these three Bonney brothers were just turning away when a burst from the ranch house caught their jet assembly, and they could only get as far as Bonneyville, thirty miles away, before they had to land. They landed right in front of the town jail.

"This Bonneyville's an awful shantytown; everybody in it is related to everybody else. The mayor, for instance, Kettle–Belly Sam Bonney, is an uncle of theirs.

"These three boys—Switchblade Joe Bonney, Jack–High Abe Bonney and Turkey–Buzzard Tom Bonney—immediately claimed sanctuary in the jail, on the grounds that they had been near to—get that; I think that indicates the line they're going to take at the trial—*near* to a political assassination. They were immediately given the protection of the jail, which is about the only well–constructed building in the place, practically a fort."

"You think that was planned in advance?" I asked.

Parros nodded emphatically. "I do. There was a hell of a big gang of these Bonneys at the jail, almost the entire able–bodied population of the place. As soon as Switchblade and Jack–High and Turkey–Buzzard landed, they were rushed inside and all the doors barred. About three minutes later, the Hickock outfit started coming in, first aircraft and then armor. They gave that town a regular Georgie Patton style blitzing."

"Yes. I'm only sorry I wasn't there to see it," Stonehenge put in. "They knocked down or burned most of the shanties, and then they went to work on the jail. The aircraft began dumping these firebombs and stun–bombs that they use to stop supercow stampedes, and the tank–guns began to punch holes in the walls. As soon as Kettle–Belly saw what he had on his hands, he radioed a call for Ranger protection. Our friend Captain Nelson went out to see what the trouble was."

"Yes. I got the story of that from Nelson," Parros put in. "Much as he hated to do it, he had to protect the Bonneys. And as soon as he'd taken a hand, Hickock had to call off his gang. But he was smart. He grabbed everything relating to the killing—the aircar and the twenty–mm auto–rifle in particular—and he's keeping them under cover. Very few people know about that, or about the fact that on physical evidence alone, he has the killing pinned on the Bonneys so well that they'll never get away with this story of being merely innocent witnesses."

"The rest, Mr. Silk, is up to us," Thrombley said. "I have Colonel Hickock's assurance that he will give us every assistance, but we simply must see to it that those creatures with the outlandish names are convicted."

I didn't have a chance to say anything to that: at that moment, one of the servants ushered Captain Nelson toward us.

"Good evening, Captain," I greeted the Ranger. "Join us, seeing that you're on foreign soil and consequently not on duty."

He sat down with us and poured a drink.

"I thought you might be interested," he said. "We gave that waiter a going-over. We wanted to know who put him up to it. He tried to sell us the line that he was a New Texan patriot, trying to kill a tyrant, but we finally got the truth out of him. He was paid a thousand pesos to do the job, by a character they call Snake–Eyes Sam Bonney. A cousin of the three who killed Mr. Cumshaw."

"Nephew of Kettle–Belly Sam," Parros interjected. "You pick him up?"

Nelson shook his head disgustedly. "He's out in the high grass somewhere. We're still looking for him. Oh, yes, and I just heard that the trial of Switchblade, and Jack–High and Turkey–Buzzard is scheduled for three days from now. You'll be notified in due form tomorrow, but I thought you might like to know in advance."

"I certainly do, and thank you, Captain…. We were just talking about you when you arrived," I mentioned. "About the arrest, or rescue, or whatever you call it, of that trio."

"Yeah. One of the jobs I'm not particularly proud of. Pity Hickock's boys didn't get hold of them before I got there. It'd of saved everybody a lot of trouble."

"Just what impression did you get at the time, Captain?" I asked. "You think Kettle–Belly knew in advance what they were going to do?"

"Sure he did. They had the whole jail fortified. Not like a jail usually is, to keep people from getting out; but like a fort, to keep people from getting in. There were no prisoners inside. I found out that they had all been released that morning."

He stopped, seemed to be weighing his words, then continued, speaking very slowly.

"Let me tell you first some things I can't testify to, couple of things that I figure went wrong with their plans.

"One of Colonel Hickock's men was on the porch to greet Mr. Cumshaw and he recognized the Bonneys. That was lucky; otherwise we might still be lookin' and wonderin' who did the shootin', which might not have been good for New Texas."

He cocked an eyebrow and I nodded. The Solar League, in similar cases, had regarded such planetary governments as due for change without notice and had promptly made the change.

"Number two," Captain Nelson continued, "that AA–shot which hit their aircar. I don't think they intended to land at the jail—it was just sort of a reserve hiding–hole. But because they'd been hit, they had to land. And they'd been slowed down so much that they couldn't dispose of the evidence before the Colonel's boys were tappin' on the door 'n' askin', couldn't they come in."

"I gather the Colonel's task–force was becoming insistent," I prompted him.

The big Ranger grinned. "Now we're on things I can testify to.

"When I got there, what had been the cell–block was on fire, and they were trying to defend the mayor's office and the warden's office. These Bonneys gave me the line that they'd been witnesses to the killing of Mr. Cumshaw by Colonel Hickock and that the Hickock outfit was trying to rub them out to keep them from testifying. I just laughed and started to walk out. Finally, they confessed that they'd shot Mr. Cumshaw, but they claimed it was right of action against political malfeasance. When they did that, I had to take them in."

"They confessed to you, before you arrested them?" I wanted to be sure of that point.

"That's right. I'm going to testify to that, Monday, when the trial is held. And that ain't all: we got their fingerprints off the car, off the gun, off some shells still in the clip, and we have the gun identified to the shells that killed Mr. Cumshaw. We got their confession fully corroborated."

I asked him if he'd give Mr. Parros a complete statement of what he'd seen and heard at Bonneyville. He was more than willing and I suggested that they go into Parros' office, where they'd be undisturbed. The Ranger and my Intelligence man got up and took a bottle of superbourbon with them. As they were leaving, Nelson turned to Hoddy, who was still with us.

"You'll have to look to your laurels, Hoddy," Nelson said. "Your Ambassador seems to be making quite a reputation for himself as a gunfighter."

"Look," Hoddy said, and though he was facing Nelson, I felt he was really talking to Stonehenge, "before I'd go up against this guy, I'd shoot myself. That way, I could be sure I'd get a nice painless job."

After they were gone, I turned to Stonehenge and Thrombley. "This seems to be a carefully prearranged killing."

They agreed.

"Then they knew *in advance* that Mr. Cumshaw would be on Colonel Hickock's front steps at about 1030. *How did they find that out?*"

"Why ... why, I'm sure I don't know," Thrombley said. It was most obvious that the idea had never occurred to him before and a side glance told me that the thought was new to Stonehenge also. "Colonel Hickock called at 0900. Mr. Cumshaw left the Embassy in an aircar a few minutes later. It took an hour and a half to fly out to the Hickock ranch...."

"I don't like the implications, Mr. Silk," Stonehenge said. "I can't believe that was how it happened. In the first place, Colonel Hickock isn't that sort of man: he doesn't use his hospitality to trap people to their death. In the second place, he wouldn't have needed to use people like these Bonneys. His own men would do anything for him. In the third place, he is one of the leaders of the annexation movement here and this was obviously an anti–annexation job. And in the fourth place—"

"Hold it!" I checked him. "Are you sure he's really on the annexation side?"

He opened his mouth to answer me quickly, then closed it, waited a moment, answered me slowly. "I can guess what you are thinking, Mr. Silk. But, remember, when Colonel Hickock came here as our first Ambassador, he came here as a man with a mission. He had studied the problem and he believed in what he came for. He has never changed.

"Let me emphasize this, sir: we know he has never changed. For our own protection, we've had to check on every real leader of the annexation movement, screening them for crackpots who might do us more harm than good. The Colonel is with us all the way.

"And now, in the fourth place, underlined by what I've just said, the Colonel and Mr. Cumshaw were really friends."

"Now you're talking!" Hoddy burst in. "I've knowed A. J. ever since I was a kid. Ever since he married old Colonel MacTodd's daughter. That just ain't the way A. J. works!"

"On the other hand, Mr. Ambassador," Thrombley said, keeping his gaze fixed on Hoddy's hands and apparently ready to both duck and shut up if Hoddy moved a finger, "you will recall, I think, that Colonel Hickock did do everything in his power to see that these Bonney brothers did not reach court alive. And, let me add," he was getting bolder, tilting his chin up a little, "it's a choice as simple as this: either Colonel Hickock told them, or we have—and this is unbelievable—a traitor in the Embassy itself."

That statement rocked even Hoddy. Even though he was probably no more than one of Natalenko's little men, he still couldn't help knowing how thoroughly we were screened, indoctrinated, and—let's face it—mind–conditioned. A traitor among us was unthinkable because we just couldn't think that way.

The silence, the sorrow, were palpable. Then I remembered, told them,

Hickock himself had been a Department man.

Stonehenge gripped his head between his hands and squeezed as if trying to bring out an idea. "All right, Mr. Ambassador, where are we now? Nobody who knew could have told the Bonney boys where Mr. Cumshaw would be at 1030, yet the three men were there waiting for him. You take it from there. I'm just a simple military man and I'm ready to go back to the simple military life as soon as possible."

I turned to Gomez. "There could be an obvious explanation. Bring us the official telescreen log. Let's see what calls were made. Maybe Mr. Cumshaw himself said something to someone that gave his destination away."

"That won't be necessary," Thrombley told me. "None of the junior clerks were on duty, and I took the only three calls that came in, myself. First, there was the call from Colonel Hickock. Then, the call about the wrist watch. And then, a couple of hours later, the call from the Hickock ranch, about Mr. Cumshaw's death."

"What was the call about the wrist watch?" I asked.

"Oh, that was from the z'Srauff Embassy," Thrombley said. "For some time, Mr. Cumshaw had been trying to get one of the very precise watches which the z'Srauff manufacture on their home planet. The z'Srauff Ambassador called, that day, to tell him that they had one for him and wanted to know when it was to be delivered. I told them the Ambassador was out, and they wanted to know where they could call him and I—"

I had never seen a man look more horror-stricken.

"Oh, my God! I'm the one who told them!"

What could I say? Not much, but I tried. "How could you know, Mr. Thrombley? You did the natural, the normal, the proper thing, on a call from one Ambassador to another."

I turned to the others, who, like me, preferred not to look at Thrombley. "They must have had a spy outside who told them the Ambassador had left the Embassy. Alone, right? And that was just what they'd been waiting for.

"But what's this about the watch, though. There's more to this than a simple favor from one Ambassador to another."

"My turn, Mr. Ambassador," Stonehenge interrupted. "Mr. Cumshaw had been trying to get one of the things at my insistence. Naval Intelligence is very much interested in them and we want a sample. The z'Srauff watches are very peculiar—they're operated by radium decay, which, of course is a universal constant. They're uniform to a tenth second and they're all synchronized with the official time at the capital city of the principal z'Srauff planet. The time used by the z'Srauff Navy."

Stonehenge deliberately paused, let that last phrase hang heavily in the air for a moment, then he continued.

"They're supposed to be used in religious observances—timing hours of prayer, I believe. They can, of course, have other uses.

"For example, I can imagine all those watches giving the wearer a light electric shock, or ringing a little bell, all over New Texas, at exactly the same moment. And then I can imagine all the z'Srauff running down into nice deep holes in the ground."

He looked at his own watch. "And that reminds me: my gang of pirates are at the spaceport by now, ready to blast off. I wonder if someone could drive me there."

"I'll drive him, boss," Hoddy volunteered. "I ain't doin' nothin' else."

I was wondering how I could break that up, plausibly and without betraying my suspicions, when Parros and Captain Nelson came out and joined us.

"I have a lot of stuff here," Parros said. "Stuff we never seemed to have noticed. For instance—"

I interrupted. "Commander Stonehenge's going to the spaceport, now," I said. "Suppose you ride with him, and brief him on what you learned, on the way. Then, when he's aboard, come back and tell us."

Hoddy looked at me for a long ten seconds. His expression started by being exasperated and ended by betraying grudging admiration.

Chapter VII

The next morning, which was Saturday, I put Thrombley in charge of the routine work of the Embassy, but first instructed him to answer all inquiries about me with the statement, literally true, that I was too immersed in work of clearing up matters left unfinished after the death of the former Ambassador for any social activities. Then I called the Hickock ranch in the west end of Sam Houston Continent, mentioning an invitation the Colonel and his daughter had extended me, and told them I would be out to see them before noon that same day. With Hoddy Ringo driving the car, I arrived about 1000, and was welcomed by Gail and her father, who had flown out the evening before, after the barbecue.

Hoddy, accompanied by a Ranger and one of Hickock's ranch hands, all three disguised in shabby and grease-stained cast-offs borrowed at the ranch, and driving a dilapidated aircar from the ranch junkyard, were sent to visit the slum village of Bonneyville. They spent all day there, posing as a trio of range tramps out of favor with the law.

I spent the day with Gail, flying over the range, visiting Hickock's herd camps and slaughtering crews. It was a pleasant day and I managed to make it constructive as well.

Because of their huge size—they ran to a live weight of around fifteen tons—and their uncertain disposition, supercows are not really domesticated. Each rancher owned the herds on his own land, chiefly by virtue of constant watchfulness over them. There were always a couple of helicopters hovering over each herd, with fast fighter planes waiting on call to come in and drop fire-bombs or stun-bombs in front of them if they showed a disposition to wander too far. Naturally, things of this size could not be shipped live to the market; they were butchered on the range, and the meat hauled out in big 'copter-trucks.

Slaughtering was dangerous and exciting work. It was done with medium tanks mounting fifty-mm guns, usually working at the rear of the herd, although a supercow herd could change directions almost in a second and the killing-tanks would then find themselves in front of a stampede. I saw several such incidents. Once Gail and I had to dive in with our car and help turn such a stampede.

We got back to the ranch house shortly before dinner. Gail went at once to change clothes; Colonel Hickock and I sat down together for a drink in his library, a beautiful room. I especially admired the walls, panelled in plastic–hardened supercow–leather.

"What do you think of our planet now, Mr. Silk?" Colonel Hickock asked.

"Well, Colonel, your final message to the State was part of the briefing I received," I replied. "I must say that I agree with your opinions. Especially with your opinion of local political practices. Politics is nothing, here, if not exciting and exacting."

"You don't understand it though." That was about half–question and half–statement. "Particularly our custom of using politicians as clay pigeons."

"Well, it is rather unusual…."

"Yes." The dryness in his tone was a paragraph of comment on my understatement. "And it's fundamental to our system of government.

"You were out all afternoon with Gail; you saw how we have to handle the supercow herds. Well, it is upon the fact that every rancher must have at his disposal a powerful force of aircraft and armor, easily convertible to military uses, that our political freedom rests. You see, our government is, in effect, an oligarchy of the big landowners and ranchers, who, in combination, have enough military power to overturn any Planetary government overnight. And, on the local level, it is a paternalistic feudalism.

"That's something that would have stood the hair of any Twentieth Century 'Liberal' on end. And it gives us the freest government anywhere in the galaxy.

"There were a number of occasions, much less frequent now than formerly, when coalitions of big ranches combined their strength and marched on the Planetary government to protect their rights from government encroachment. This sort of thing could only be resorted to in defense of some inherent right, and never to infringe on the rights of others. Because, in the latter case, other armed coalitions would have arisen, as they did once or twice during the first three decades of New Texan history, to resist.

"So the right of armed intervention by the people when the government invaded or threatened their rights became an acknowledged part of our political system.

"And—this arises as a natural consequence—you can't give a man with five hundred employees and a force of tanks and aircraft the right to resist the government, then at the same time deny that right to a man who has only his own pistol or machete."

"I notice the President and the other officials have themselves surrounded by guards to protect them from individual attack," I said. "Why doesn't the government, as such, protect itself with an army and air force large enough to resist any possible coalition of the big ranchers?"

"*Because we won't let the government get that strong!*" the Colonel said forcefully. "That's one of the basic premises. We have no standing army, only the New Texas Rangers. And the legislature won't authorize any standing army, or appropriate funds to support one. Any member of the legislature who tried it would get what Austin Maverick got, a couple of weeks ago, or what Sam Saltkin got, eight years ago, when he proposed a law for the compulsory registration and licensing of firearms. The opposition to that tax scheme of Maverick's wasn't because of what it would cost the public in taxes, but from fear of what the government could do with the money after they got it.

"Keep a government poor and weak and it's your servant; let it get rich and powerful and it's your master. We don't want any masters here on New Texas."

"But the President has a bodyguard," I noted.

"Casualty rate was too high," Hickock explained. "Remember, the President's job is inherently impossible: he has to represent *all* the people."

I thought that over, could see the illogical logic, but … "How about your rancher oligarchy?"

He laughed. "Son, if I started acting like a master around this ranch in the morning, they'd find my body in an irrigation ditch before sunset.

"Sure, if you have a real army, you can keep the men under your thumb—use one regiment or one division to put down mutiny in another. But when you have only five hundred men, all of whom know everybody else and all of them armed, you just act real considerate of them if you want to keep on living."

"Then would you say that the opposition to annexation comes from the people who are afraid that if New Texas enters the Solar League, there will be League troops sent here and this ... this interesting system of insuring government responsibility to the public would be brought to an end?"

"Yes. If you can show the people of this planet that the League won't interfere with local political practices, you'll have a 99.95 percent majority in favor of annexation. We're too close to the z'Srauff star–cluster, out here, not to see the benefits of joining the Solar League."

We left the Hickock ranch on Sunday afternoon and while Hoddy guided our air–car back to New Austin, I had a little time to revise some of my ideas about New Texas. That is, I had time to think during those few moments when Hoddy wasn't taking advantage of our diplomatic immunity to invent new air–ground traffic laws.

My thoughts alternated between the pleasure of remembering Gail's gay company and the gloom of understanding the complete implications of the Colonel's clarifying lectures. Against the background of his remarks, I could find myself appreciating the Ghopal–Klüng–Natalenko reasoning: the only way to cut the Gordian knot was to have another Solar League Ambassador killed.

And, whenever I could escape thinking about the fact that the next Ambassador to be the clay pigeon was me, I found myself wondering if I wanted the League to take over. Annexation, yes; New Texas customs would be protected under a treaty of annexation. But the "justified conquest" urged by Machiavelli, Jr.? No.

I was still struggling with the problem when we reached the Embassy about 1700. Everyone was there, including Stonehenge, who had returned two hours earlier with the good news that the fleet had moved into position only sixty light-minutes off Capella IV. I had reached the point in my thinking where I had decided it was useless to keep Hoddy and Stonehenge apart except as an exercise in mental agility. Inasmuch as my brain was already weight-lifting, swinging from a flying trapeze to elusive flying rings while doing triple somersaults and at the same time juggling seven Indian clubs, I skipped the whole matter.

But I'm fairly certain that it wasn't till then that Hoddy had a chance to deliver his letter-of-credence to Stonehenge.

After dinner, we gathered in my office for our coffee and a final conference before the opening of the trial the next morning.

Stonehenge spoke first, looking around the table at everyone except me.

"No matter what happens, we have the fleet within call. Sir Rodney's been active picking up those z'Srauff meteor-mining boats. They no longer have a tight screen around the system. We do. I don't think that anyone, except us, knows that the fleet's where it is."

No matter what happens, I thought glumly, and the phrase explained why he hadn't been able to look at me.

"Well, boss, I gave you my end of it, comin' in," Hoddy said. "Want me to go over it again? All right. In Bonneyville, we found half a dozen people who can swear that Kettle-Belly Sam Bonney was making preparations to protect those three brothers an hour before Ambassador Cumshaw was shot. The whole town's sorer than hell at Kettle-Belly for antagonizing the Hickock outfit and getting the place shot up the way it was. And we have witnesses that Kettle-Belly was in some kind of deal with the z'Srauff, too. The Rangers gathered up eight of them, who can swear to the preparations and to the fact that Kettle-Belly had z'Srauff visitors on different occasions before the shooting."

"That's what we want," Stonehenge said. "Something that'll connect this murder with the z'Srauff."

"Well, wait till you hear what I've got," Parros told him. 'In the first place, we traced the gun and the air-car. The Bonney brothers bought them both from z'Srauff merchants, for ridiculously nominal prices. The merchant who sold the aircar is normally in the dry-goods business, and the one who sold the auto–rifle runs a toy shop. In their whole lives, those three boys never had enough money among them to pay the list price of the gun, let alone the car. That is, not until a week before the murder."

"They got prosperous, all of a sudden?" I asked.

"Yes. Two weeks before the shooting, Kettle–Belly Sam's bank account got a sudden transfusion: some anonymous benefactor deposited 250,000 pesos—about a hundred thousand dollars—to his credit. He drew out 75,000 of it and some of the money turned up again in the hands of Switchblade and Jack–High and Turkey–Buzzard. Then, a week before you landed here, he got another hundred thousand from the same anonymous source and he drew out twenty thousand of that. We think that was the money that went to pay for the attempted knife–job on Hutchinson. Two days before the barbecue, the waiter deposited a thousand at the New Austin Packers' and Shippers' Trust."

"Can you get that introduced as evidence at the trial?" I asked.

"Sure. Kettle–Belly banks at a town called Crooked Creek, about forty miles from Bonneyville. We have witnesses from the bank.

"I also got the dope on the line the Bonney brothers are going to take at the trial. They have a lawyer, Clement A. Sidney, a member of what passes for the Socialist Party on this planet. The defense will take the line of full denial of everything. The Bonneys are just three poor but honest boys who are being framed by the corrupt tools of the Big Ranching Interests."

Hoddy made an impolite noise. "Whatta we got to worry about, then?" he demanded. "They're a cinch for conviction."

"I agree with that," Stonehenge said. "If they tried to base their defense on political conviction and opposition by the Solar League, they might have a chance. This way, they haven't."

"All right, gentlemen," I said, "I take it that we're agreed that we must all follow a single line of policy and not work at cross–purposes to each other?"

They all agreed to that instantly, but with a questioning note in their voices.

"Well, then, I trust you all realize that we cannot, under any circumstances, allow those three brothers to be convicted in this court," I added.

There was a moment of startled silence, while Hoddy and Stonehenge and Parros and Thrombley were understanding what they had just heard. Then Stonehenge cleared his throat and said:

"Mr. Ambassador! I'm sure that you have some excellent reasons for that remarkable statement, but I must say—"

"It was a really colossal error on somebody's part," I said, "that this case was allowed to get into the Court of Political Justice. It never should have. And if we take a part in the prosecution, or allow those men to be convicted, we will establish a precedent to support the principle that a foreign Ambassador is, on this planet, defined as a practicing local politician.

"I will invite you to digest that for a moment."

A moment was all they needed. Thrombley was horrified and dithered incoherently. Stonehenge frowned and fidgeted with some papers in front of him. I could see several thoughts gathering behind his eyes, including, I was sure, a new view of his instructions from Klüng.

Even Hoddy got at least part of it. "Why, that means that anybody can bump off any diplomat he doesn't like...." he began.

"That is only part of it, Mr. Ringo," Thrombley told him. "It also means that a diplomat, instead of being regarded as the representative of his own government, becomes, in effect, a functionary of the government of New Texas. Why, all sorts of complications could arise...."

"It certainly would impair, shall we say, the principle of extraterritoriality of Embassies," Stonehenge picked it up. "And it would practically destroy the principle of diplomatic immunity."

"Migawd!" Hoddy looked around nervously, as though he could already hear an army of New Texas Rangers, each with a warrant for Hoddy Ringo, battering at the gates.

"We'll have to do something!" Gomez, the Secretary of the Embassy, said.

"I don't know what," Stonehenge said. "The obvious solution would be, of course, to bring charges against those Bonney Boys on simple first–degree murder, which would be tried in an ordinary criminal court. But it's too late for that now. We wouldn't have time to prevent their being arraigned in this Political Justice court, and once a defendant is brought into court, on this planet, he cannot be brought into court again for the same act. Not the same *crime*, the same *act*."

I had been thinking about this and I was ready. "Look, we must bring those Bonney brothers to trial. It's the only effective way of demonstrating to the public the simple fact that Ambassador Cumshaw was murdered at the instigation of the z'Srauff. We dare not allow them to be convicted in the Court of Political Justice, for the reasons already stated. And to maintain the prestige of the Solar League, we dare not allow them to go unpunished."

"We can have it one way," Parros said, "and maybe we can have it two ways. But I'm damned if I can see how we can have it all three ways."

I wasn't surprised that he didn't see it; he hadn't had the same urgency goading him which had forced me to find the answer. It wasn't an answer that I liked, but I was in the position where I had no choice.

"Well, here's what we have to do, gentlemen," I began, and from the respectful way they regarded me, from the attention they were giving my words, I got a sudden thrill of pride. For the first time since my scrambled arrival, I was really *Ambassador* Stephen Silk.

Chapter VIII

A couple of New Texas Ranger tanks met the Embassy car four blocks from the Statehouse and convoyed us into the central plaza, where the barbecue had been held on the Friday afternoon that I had arrived on New Texas. There was almost as dense a crowd as the last time I had seen the place; but they were quieter, to the extent that there were no bands, and no shooting, no cowbells or whistles. The barbecue pits were going again, however, and hawkers were pushing or propelling their little wagons about, vending sandwiches. I saw a half a dozen big twenty-foot teleview screens, apparently wired from the courtroom.

As soon as the Embassy car and its escorting tanks reached the plaza, an ovation broke out. I was cheered, with the high-pitched *yipeee!* of New Texans and adjured and implored not to let them so-and-sos get away with it.

There was a veritable army of Rangers on guard at the doors of the courtroom. The only spectators being admitted to the courtroom seemed to be prominent citizens with enough pull to secure passes.

Inside, some of the spectators' benches had been removed to clear the front of the room. In the cleared space, there was one bulky shape under a cloth cover that seemed to be the air-car and another cloth-covered shape that looked like a fifty-mm dual-purpose gun. Smaller exhibits, including a twenty-mm auto-rifle, were piled on the friends-of-the-court table. The prosecution table was already occupied—Colonel Hickock, who waved a greeting to me, three or four men who looked like well-to-do ranchers, and a delegation of lawyers.

"Samuel Goodham," Parros, beside me, whispered, indicating a big, heavy-set man with white hair, dressed in a dark suit of the cut that had been fashionable on Terra seventy-five years ago. "Best criminal lawyer on the planet. Hickock must have hired him."

There was quite a swarm at the center table, too. Some of them were ranchers, a couple in aggressively shabby workclothes, and there were several members of the Diplomatic Corps. I shook hands with them and gathered that they, like myself, were worried about the precedent that might be established by this trial. While I was introducing Hoddy Ringo as my attaché extraordinary, which was no less than the truth, the defense party came in.

There were only three lawyers—a little, rodent–faced fellow, whom Parros pointed out as Clement Sidney, and two assistants. And, guarded by a Ranger and a couple of court–bailiffs, the three defendants, Switchblade Joe, Jack–High Abe and Turkey–Buzzard Tom Bonney. There was probably a year or so age different from one to another, but they certainly had a common parentage. They all had pale eyes and narrow, loose–lipped faces. Subnormal and probably psychopathic, I thought. Jack–High Abe had his left arm in a sling and his left shoulder in a plaster cast. The buzz of conversation among the spectators altered its tone subtly and took on a note of hostility as they entered and seated themselves.

The balcony seemed to be crowded with press representatives. Several telecast cameras and sound pickups had been rigged to cover the front of the room from various angles, a feature that had been missing from the trial I had seen with Gail on Friday.

Then the judges entered from a door behind the bench, which must have opened from a passageway under the plaza, and the court was called to order.

The President Judge was the same Nelson who had presided at the Whately trial and the first thing on the agenda seemed to be the selection of a new board of associate judges. Parros explained in a whisper that the board which had served on the previous trial would sit until that could be done.

A slip of paper was drawn from a box and a name was called. A man sitting on one of the front rows of spectators' seats got up and came forward. One of Sidney's assistants rummaged through a card file he had in front of him and handed a card to the chief of the defense. At once, Sidney was on his feet.

"Challenged, for cause!" he called out. "This man is known to have declared, in conversation at the bar of the Silver Peso Saloon, here in New Austin, that these three boys, my clients, ought all to be hanged higher than Haman."

"Yes, I said that!" the venireman declared. "I'll repeat it right here: all three of these murdering skunks ought to be hanged higher than—"

"Your Honor!" Sidney almost screamed. "If, after hearing this man's brazen declaration of bigoted class hatred against my clients, he is allowed to sit on that bench—"

Judge Nelson pounded with his gavel. "You don't have to instruct me in my judicial duties, Counselor," he said. "The venireman has obviously disqualified himself by giving evidence of prejudice. Next name."

The next man was challenged: he was a retired packing-house operator in New Austin, and had once expressed the opinion that Bonneyville and everybody in it ought to be H-bombed off the face of New Texas.

This Sidney seemed to have gotten the name of everybody likely to be called for court duty and had something on each one of them, because he went on like that all morning.

"You know what I think," Stonehenge whispered to me, leaning over behind Parros. "I think he's just stalling to keep the court in session until the z'Srauff fleet gets here. I wish we could get hold of one of those wrist watches."

"I can get you one, before evening," Hoddy offered, "if you don't care what happens to the mutt that's wearin' it."

"Better not," I decided. "Might tip them off to what we suspect. And we don't really need one: Sir Rodney will have patrols out far enough to get warning in time."

We took an hour, at noon, for lunch, and then it began again. By 1647, fifteen minutes before court should be adjourned, Judge Nelson ordered the bailiff to turn the clock back to 1300. The clock was turned back again when it reached 1645. By this time, Clement Sidney was probably the most unpopular man on New Texas.

Finally, Colonel Andrew J. Hickock rose to his feet.

"Your Honor: the present court is not obliged to retire from the bench until another court has been chosen as they are now sitting as a court in being. I propose that the trial begin, with the present court on the bench."

Sidney began yelling protests. Hoddy Ringo pulled his neckerchief around under his left ear and held the ends above his head. Nanadabadian, the Ambassador from Beta Cephus IV, drew his biggest knife and began trying the edge on a sheet of paper.

"Well, Your Honor, I certainly do not wish to act in an obstructionist manner. The defense agrees to accept the present court," Sidney decided.

"Prosecution agrees to accept the present court," Goodham parroted.

"The present court will continue on the bench, to try the case of the Friends of Silas Cumshaw, deceased, versus Switchblade Joe Bonney, Jack–High Abe Bonney, Turkey–Buzzard Tom Bonney, et als." Judge Nelson rapped with his gavel. "Court is herewith adjourned until 0900 tomorrow."

Chapter IX

The trial got started the next morning with a minimum amount of objections from Sidney. The charges and specifications were duly read, the three defendants pleaded not guilty, and then Goodham advanced with a paper in his hand to address the court. Sidney scampered up to take his position beside him.

"Your Honor, the prosecution wishes, subject to agreement of the defense, to enter the following stipulations, to wit: First, that the late Silas Cumshaw was a practicing politician within the meaning of the law. Second, that he is now dead, and came to his death in the manner attested to by the coroner of Sam Houston Continent. Third, that he came to his death at the hands of the defendants here present."

In all my planning, I'd forgotten that. I couldn't let those stipulations stand without protest, and at the same time, if I protested the characterization of Cumshaw as a practicing politician, the trial could easily end right there. So I prayed for a miracle, and Clement Sidney promptly obliged me.

"Defense won't stipulate anything!" he barked. "My clients, here, are victims of a monstrous conspiracy, a conspiracy to conceal the true facts of the death of Silas Cumshaw. They ought never to have been arrested or brought here, and if the prosecution wants to establish anything, they can do it by testimony, in the regular and lawful way. This practice of free-wheeling stipulation is only one of the many devices by which the courts of this planet are being perverted to serve the corrupt and unjust ends of a gang of reactionary landowners!"

Judge Nelson's gavel hit the bench with a crack like a rifle shot.

"Mr. Sidney! In justice to your clients, I would hate to force them to change lawyers in the middle of their trial, but if I hear another remark like that about the courts of New Texas, that's exactly what will happen, because you'll be in jail for contempt! Is that clear, Mr. Sidney?"

I settled back with a deep sigh of relief which got me, I noticed, curious stares from my fellow Ambassadors. I disregarded the questions in their glances; I had what I wanted.

They began calling up the witnesses.

First, the doctor who had certified Ambassador Cumshaw's death. He gave a concise description of the wounds which had killed my predecessor. Sidney was trying to make something out of the fact that he was Hickock's family physician, and consuming more time, when I got up.

"Your Honor, I am present here as *amicus curiae*, because of the obvious interest which the Government of the Solar League has in this case...."

"Objection!" Sidney yelled.

"Please state it," Nelson invited.

"This is a court of the people of the planet of New Texas. This foreign emissary of the Solar League, sent here to conspire with New Texan traitors to the end that New Texans shall be reduced to a supine and ravished satrapy of the all–devouring empire of the Galaxy—"

Judge Nelson rapped sharply.

"Friends of the court are defined as persons having a proper interest in the case. As this case arises from the death of the former Ambassador of the Solar League, I cannot see how the present Ambassador and his staff can be excluded. Overruled." He nodded to me. "Continue, Mr. Ambassador."

"As I understand, I have the same rights of cross–examination of witnesses as counsel for the prosecution and defense; is that correct, Your Honor?" It was, so I turned to the witness. "I suppose, Doctor, that you have had quite a bit of experience, in your practice, with gunshot wounds?"

He chuckled. "Mr. Ambassador, it is gunshot–wound cases which keep the practice of medicine and surgery alive on this planet. Yes, I definitely have."

"Now, you say that the deceased was hit by six different projectiles: right shoulder almost completely severed, right lung and right ribs blown out of the chest, spleen and kidneys so intermingled as to be practically one, and left leg severed by complete shattering of the left pelvis and hip–joint?"

"That's right."

I picked up the 20–mm auto–rifle—it weighed a good sixty pounds—from the table, and asked him if this weapon could have inflicted such wounds. He agreed that it both could and had.

"This the usual type of weapon used in your New Texas political liquidations?" I asked.

"Certainly not. The usual weapons are pistols; sometimes a hunting–rifle or a shotgun."

I asked the same question when I cross–examined the ballistics witness.

"Is this the usual type of weapon used in your New Texas political liquidations?"

"No, not at all. That's a very expensive weapon, Mr. Ambassador. Wasn't even manufactured on this planet; made by the z'Srauff star–cluster. A weapon like that sells for five, six hundred pesos. It's used for shooting really big game—supermastodon, and things like that. And, of course, for combat."

"It seems," I remarked, "that the defense is overlooking an obvious point there. I doubt if these three defendants ever, in all their lives, had among them the price of such a weapon."

That, of course, brought Sidney to his feet, sputtering objections to this attempt to disparage the honest poverty of his clients, which only helped to call attention to the point.

Then the prosecution called in a witness named David Crockett Longfellow. I'd met him at the Hickock ranch; he was Hickock's butler. He limped from an old injury which had retired him from work on the range. He was sworn in and testified to his name and occupation.

"Do you know these three defendants?" Goodham asked him.

"Yeah. I even marked one of them for future identification," Longfellow replied.

Sidney was up at once, shouting objections. After he was quieted down, Goodham remarked that he'd come to that point later, and began a line of questioning to establish that Longfellow had been on the Hickock ranch on the day when Silas Cumshaw was killed.

"Now," Goodham said, "will you relate to the court the matters of interest which came to your personal observation on that day."

Longfellow began his story. "At about 0900, I was dustin' up and straightenin' things in the library while the Colonel was at his desk. All of a sudden, he said to me, 'Davy, suppose you call the Solar Embassy and see if Mr. Cumshaw is doin' anything today; if he isn't, ask him if he wants to come out.' I was workin' right beside the telescreen. So I called the Solar League Embassy. Mr. Thrombley took the call, and I asked him was Mr. Cumshaw around. By this time, the Colonel got through with what he was doin' at the desk and came over to the screen. I went back to my work, but I heard the Colonel askin' Mr. Cumshaw could he come out for the day, an' Mr. Cumshaw sayin', yes, he could; he'd be out by about 1030.

"Well, 'long about 1030, his air–car came in and landed on the drive. Little single–seat job that he drove himself. He landed it about a hundred feet from the outside veranda, like he usually did, and got out.

"Then, this other car came droppin' in from outa nowhere. I didn't pay it much attention; thought it might be one of the other Ambassadors that Mr. Cumshaw'd brung along. But Mr. Cumshaw turned around and looked at it, and then he started to run for the veranda. I was standin' in the doorway when I seen him startin' to run. I jumped out on the porch, quick–like, and pulled my gun, and then this auto–rifle begun firin' outa the other car. There was only eight or ten shots fired from this car, but most of them hit Mr. Cumshaw."

Goodham waited a few moments. Longfellow's voice had choked and there was a twitching about his face, as though he were trying to suppress tears.

"Now, Mr. Longfellow," Goodham said, "did you recognize the people who were in the car from which the shots came?"

"Yeah. Like I said, I cut a mark on one of them. That one there: Jack–High Abe Bonney. He was handlin' the gun, and from where I was, he had his left side to me. I was try_n' for his head, but I always overshoot, so I have the habit of holdin' low. This time I held too low." He looked at Jack–High in coldly poisonous hatred. "I'll be sorry about that as long as I live."

"And who else was in the car?"

"The other two curs outa the same litter: Switchblade an' Turkey–Buzzard, over there."

Further questioning revealed that Longfellow had had no direct knowledge of the pursuit, or the siege of the jail in Bonneyville. Colonel Hickock had taken personal command of that, and had left Longfellow behind to call the Solar League Embassy and the Rangers. He had made no attempt to move the body, but had left it lying in the driveway until the doctor and the Rangers arrived.

Goodham went to the middle table and picked up a heavy automatic pistol.

"I call the court's attention to this pistol. It is an eleven–mm automatic, manufactured by the Colt Firearms Company of New Texas, a licensed subsidiary of the Colt Firearms Company of Terra." He handed it to Longfellow. "Do you know this pistol?" he asked.

Longfellow was almost insulted by the question. Of course he knew his own pistol. He recited the serial number, and pointed to different scars and scratches on the weapon, telling how they had been acquired.

"The court accepts that Mr. Longfellow knows his own weapon," Nelson said. "I assume that this is the weapon with which you claim to have shot Jack–High Abe Bonney?"

It was, although Longfellow resented the qualification.

"That's all. Your witness, Mr. Sidney," Goodham said.

Sidney began an immediate attack.

Questioning Longfellow's eyesight, intelligence, honesty and integrity, he tried to show personal enmity toward the Bonneys. He implied that Longfellow had been conspiring with Cumshaw to bring about the conquest of New Texas by the Solar League. The verbal exchange became so heated that both witness and attorney had to be admonished repeatedly from the bench. But at no point did Sidney shake Longfellow from his one fundamental statement, that the Bonney brothers had shot Silas Cumshaw and that he had shot Jack–High Abe Bonney in the shoulder.

When he was finished, I got up and took over.

"Mr. Longfellow, you say that Mr. Thrombley answered the screen at the Solar League Embassy," I began. "You know Mr. Thrombley?"

"Sure, Mr. Silk. He's been out at the ranch with Mr. Cumshaw a lotta times."

"Well, beside yourself and Colonel Hickock and Mr. Cumshaw and, possibly, Mr. Thrombley, who else knew that Mr. Cumshaw would be at the ranch at 1030 on that morning?"

Nobody. But the aircar had obviously been waiting for Mr. Cumshaw; the Bonneys must have had advance knowledge. My questions made that point clear despite the obvious—and reluctantly court–sustained—objections from Mr. Sidney.

"That will be all, Mr. Longfellow; thank you. Any questions from anybody else?"

There being none, Longfellow stepped down. It was then a few minutes before noon, so Judge Nelson recessed court for an hour and a half.

In the afternoon, the surgeon who had treated Jack–High Abe Bonney's wounded shoulder testified, identifying the bullet which had been extracted from Bonney's shoulder. A ballistics man from Ranger crime–lab followed him to the stand and testified that it had been fired from Longfellow's Colt. Then Ranger Captain Nelson took the stand. His testimony was about what he had given me at the Embassy, with the exception that the Bonneys' admission that they had shot Ambassador Cumshaw was ruled out as having been made under duress.

However, Captain Nelson's testimony didn't need the confessions.

The cover was stripped off the air–car, and a couple of men with a power–dolly dragged it out in front of the bench. The Ranger Captain identified it as the car which he had found at the Bonneyville jail. He went over it with an ultra-violet flashlight and showed where he had written his name and the date on it with fluorescent ink. The effects of AA-fire were plainly evident on it.

Then the other shrouded object was unveiled and identified as the gun which had disabled the air-car. Colonel Hickock identified the gun as the one with which he had fired on the air-car. Finally, the ballistics expert was brought back to the stand again, to link the two by means of fragments found in the car.

Then Goodham brought Kettle-Belly Sam Bonney to the stand.

The Mayor of Bonneyville was a man of fifty or so, short, partially bald, dressed in faded blue Levis, a frayed white shirt, and a grease-spotted vest. There was absolutely no mystery about how he had acquired his nickname. He disgorged a cud of tobacco into a spittoon, took the oath with unctuous solemnity, then reloaded himself with another chew and told his version of the attack on the jail.

At about 1045 on the day in question, he testified, he had been in his office, hard at work in the public service, when an air-car, partially disabled by gunfire, had landed in the street outside and the three defendants had rushed in, claiming sanctuary. From then on, the story flowed along smoothly, following the lines predicted by Captain Nelson and Parros. Of course he had given the fugitives shelter; they had claimed to have been near to a political assassination and were in fear of their lives.

Under Sidney's cross-examination, and coaching, he poured out the story of Bonneyville's wrongs at the hands of the reactionary landowners, and the atrocious behavior of the Hickock goon-gang. Finally, after extracting the last drop of class-hatred venom out of him, Sidney turned him over to me.

"How many men were inside the jail when the three defendants came claiming sanctuary?" I asked.

He couldn't rightly say, maybe four or five.

"Closer twenty-five, according to the Rangers. How many of them were prisoners in the jail?"

"Well, none. The prisoners was all turned out that mornin'. They was just common drunks, disorderly conduct cases, that kinda thing. We turned them out so's we could make some repairs."

"You turned them out because you expected to have to defend the jail; because you knew in advance that these three would be along claiming sanctuary, and that Colonel Hickock's ranch hands would be right on their heels, didn't you?" I demanded.

It took a good five minutes before Sidney stopped shouting long enough for Judge Nelson to sustain the objection.

"You knew these young men all their lives, I take it. What did you know about their financial circumstances, for instance?"

"Well, they've been ground down an' kept poor by the big ranchers an' the money–guys...."

"Then weren't you surprised to see them driving such an expensive aircar?"

"I don't know as it's such an expensive—" he shut his mouth suddenly.

"You know where they got the money to buy that car?" I pressed.

Kettle–Belly Sam didn't answer.

"From the man who paid them to murder Ambassador Silas Cumshaw?" I kept pressing. "Do you know how much they were paid for that job? Do you know where the money came from? Do you know who the go–between was, and how much he got, and how much he kept for himself? Was it the same source that paid for the recent attempt on President Hutchinson's life?"

"I refuse to answer!" the witness declared, trying to shove his chest out about half as far as his midriff. "On the grounds that it might incriminate or degrade me!"

"You can't degrade a Bonney!" a voice from the balcony put in.

"So then," I replied to the voice, "what he means is, incriminate." I turned to the witness. "That will be all. Excused."

As Bonney left the stand and was led out the side door, Goodham addressed the bench.

"Now, Your Honor," he said, "I believe that the prosecution has succeeded in definitely establishing that these three defendants actually did fire the shot which, on April 22, 2193, deprived Silas Cumshaw of his life. We will now undertake to prove...."

Followed a long succession of witnesses, each testifying to some public or private act of philanthropy, some noble trait of character. It was the sort of thing which the defense lawyer in the Whately case had been so willing to stipulate. Sidney, of course, tried to make it all out to be part of a sinister conspiracy to establish a Solar League fifth column on New Texas. Finally, the prosecution rested its case.

I entertained Gail and her father at the Embassy, that evening. The street outside was crowded with New Texans, all of them on our side, shouting slogans like, "Death to the Bonneys!" and "Vengeance for Cumshaw!" and "Annexation Now!" Some of it was entirely spontaneous, too. The Hickocks, father and daughter, were given a tremendous ovation, when they finally left, and followed to their hotel by cheering crowds. I saw one big banner, lettered: 'DON'T LET NEW TEXAS GO TO THE DOGS.' and bearing a crude picture of a z'Srauff. I seemed to recall having seen a couple of our Marines making that banner the evening before in the Embassy patio, but....

Chapter X

The next morning, the third of the trial, opened with the defense witnesses, character–witnesses for the three killers and witnesses to the political iniquities of Silas Cumshaw.

Neither Goodham nor I bothered to cross–examine the former. I couldn't see how any lawyer as shrewd as Sidney had shown himself to be would even dream of getting such an array of thugs, cutthroats, sluts and slatterns into court as character witnesses for anybody.

The latter, on the other hand, we went after unmercifully, revealing, under their enmity for Cumshaw, a small, hard core of bigoted xenophobia and selfish fear. Goodham did a beautiful job on that; he seemed able, at a glance, to divine exactly what each witness's motivation was, and able to make him or her betray that motivation in its least admirable terms. Finally the defense rested, about a quarter–hour before noon.

I rose and addressed the court:

"Your Honor, while both the prosecution and the defense have done an admirable job in bringing out the essential facts of how my predecessor met his death, there are many features about this case which are far from clear to me. They will be even less clear to my government, which is composed of men who have never set foot on this planet. For this reason, I wish to call, or recall, certain witnesses to clarify these points."

Sidney, who had begun shouting objections as soon as I had gotten to my feet, finally managed to get himself recognized by the court.

"This Solar League Ambassador, Your Honor, is simply trying to use the courts of the Planet of New Texas as a sounding–board for his imperialistic government's propaganda...."

"You may reassure yourself, Mr. Sidney," Judge Nelson said. "This court will not allow itself to be improperly used, or improperly swayed, by the Ambassador of the Solar League. This court is interested only in determining the facts regarding the case before it. You may call your witnesses, Mr. Ambassador." He glanced at his watch. "Court will now recess for an hour and a half; can you have them here by 1330?"

I assured him I could after glancing across the room at Ranger Captain Nelson and catching his nod.

My first witness, that afternoon was Thrombley. After the formalities of getting his name and connection with the Solar League Embassy on the record, I asked him, "Mr. Thrombley, did you, on the morning of April 22, receive a call from the Hickock ranch for Mr. Cumshaw?'

"Yes, indeed, Mr. Ambassador. The call was from Mr. Longfellow, Colonel Hickock's butler. He asked if Mr. Cumshaw were available. It happened that Mr. Cumshaw was in the same room with me, and he came directly to the screen. Then Colonel Hickock appeared in the screen, and inquired if Mr. Cumshaw could come out to the ranch for the day; he said something about superdove shooting."

"You heard Mr. Cumshaw tell Colonel Hickock that he would be out at the ranch at about 1030?" Thrombley said he had. "And, to your knowledge, did anybody else at the Embassy hear that?"

"Oh, no, sir; we were in the Ambassador's private office, and the screen there is tap–proof."

"And what other calls did you receive, prior to Mr. Cumshaw's death?"

"About fifteen minutes after Mr. Cumshaw had left, the z'Srauff Ambassador called, about a personal matter. As he was most anxious to contact Mr. Cumshaw, I told him where he had gone."

"Then, to your knowledge, outside of yourself, Colonel Hickock, and his butler, the z'Srauff Ambassador was the only person who could have known that Mr. Cumshaw's car would be landing on Colonel Hickock's drive at or about 1030. Is that correct?"

"Yes, plus anybody whom the z'Srauff Ambassador might have told."

"Exactly!" I pounced. Then I turned and gave the three Bonney brothers a sweeping glance. "Plus anybody the z'Srauff Ambassador might have told.... That's all. Your witness, Mr. Sidney."

Sidney got up, started toward the witness stand, and then thought better of it.

"No questions," he said.

The next witness was a Mr. James Finnegan; he was identified as cashier of the Crooked Creek National Bank. I asked him if Kettle–Belly Sam Bonney did business at his bank; he said yes.

"Anything unusual about Mayor Bonney's account?" I asked.

"Well, it's been unusually active lately. Ordinarily, he carries around two–three thousand pesos, but about the first of April, that took a big jump. Quite a big jump; two hundred and fifty thousand pesos, all in a lump."

"When did Kettle–Belly Sam deposit this large sum?" I asked.

"He didn't. The money came to us in a cashier's check on the Ranchers' Trust Company of New Austin with an anonymous letter asking that it be deposited to Mayor Bonney's account. The letter was typed on a sheet of yellow paper in Basic English."

"Do you have that letter now?" I asked.

"No, I don't. After we'd recorded the new balance, Kettle–Belly came storming in, raising hell because we'd recorded it. He told me that if we ever got another deposit like that, we were to turn it over to him in cash. Then he wanted to see the letter, and when I gave it to him, he took it over to a telescreen booth, and drew the curtains. I got a little busy with some other matters, and the next time I looked, Kettle–Belly was gone and some girl was using the booth."

"That's very interesting, Mr. Finnegan. Was that the last of your unusual business with Mayor Bonney?"

"Oh, no. Then, about two weeks before Mr. Cumshaw was killed, Kettle–Belly came in and wanted 50,000 pesos, in a big hurry, in small bills. I gave it to him, and he grabbed at the money like a starved dog at a bone, and upset a bottle of red perma–ink, the sort we use to refill our bank seals.

Three of the bills got splashed. I offered to exchange them, but he said, 'Hell with it; I'm in a hurry,' and went out. The next day, Switchblade Joe Bonney came in to make payment on a note we were holding on him. He used those three bills in the payment.

"Then, about a week ago, there was another cashier's check came in for Kettle–Belly. This time, there was no letter; just one of our regular deposit–slips. No name of depositor. I held the check, and gave it to Kettle–Belly. I remember, when it came in, I said to one of the clerks, 'Well, I wonder who's going to get bumped off this time.' And sure enough …"

Sidney's yell of, "Objection!" was all his previous objections gathered into one.

"You say the letter accompanying the first deposit, the one in Basic English, was apparently taken away by Kettle–Belly Sam Bonney. If you saw another letter of the same sort, would you be able to say whether or not it might be like the one you mentioned?"

Sidney vociferating more objections; I was trying to get expert testimony without previous qualification….

"Not at all, Mr. Sidney,' Judge Nelson ruled. "Mr. Silk has merely asked if Mr. Finnegan could say whether one document bore any resemblance to another."

I asked permission to have another witness sworn in while Finnegan was still on the stand, and called in a Mr. Boone, the cashier of the Packers' and Brokers' Trust Company of New Austin. He had with him a letter, typed on yellow paper, which he said had accompanied an anonymous deposit of two hundred thousand pesos. Mr. Finnegan said that it was exactly like the one he had received, in typing, grammar and wording, all but the name of the person to whose account the money was to be deposited.

"And whose account received this anonymous benefaction, Mr. Boone?" I asked.

"The account," Boone replied, "of Mr. Clement Sidney."

I was surprised that Judge Nelson didn't break the handle of his gavel, after that. Finally, after a couple of threats to clear the court, order was restored. Mr. Sidney had no questions to ask this time, either.

The bailiff looked at the next slip of paper I gave him, frowned over it, and finally asked the court for assistance.

"I can't pronounce this—here thing, at all," he complained.

One of the judges finally got out a mouthful of growls and yaps, and gave it to the clerk of the court to copy into the record. The next witness was a z'Srauff, and in the New Texan garb he was wearing, he was something to open my eyes, even after years on the Hooligan Diplomats.

After he took the stand, the clerk of the court looked at him blankly for a moment. Then he turned to Judge Nelson.

"Your Honor, how am I gonna go about swearing him in?" he asked. "What does a z'Srauff swear by, that's binding?"

The President Judge frowned for a moment. "Does anybody here know Basic well enough to translate the oath?" he asked.

"I think I can," I offered. "I spent a great many years in our Consular Service, before I was sent here. We use Basic with a great many alien peoples."

"Administer the oath, then," Nelson told me.

"Put up right hand," I told the z'Srauff. "Do you truly say, in front of Great One who made all worlds, who has knowledge of what is in the hearts of all persons, that what you will say here will be true, all true, and not anything that is not true, and will you so say again at time when all worlds end? Do you so truly say?"

"Yes. I so truly say."

"Say your name."

"Ppmegll Kkuvtmmecc Cicici."

"What is your business?"

"I put things made of cloth into this world, and I take meat out of this world."

"Where do you have your house?"

"Here in New Austin, over my house of business, on Coronado Street."

"What people do you see in this place that you have made business with?"

Ppmegll Kkuvtmmecc Cicici pointed a three-fingered hand at the Bonney brothers.

"What business did you make with them?"

"I gave them for money a machine which goes on the ground and goes in the air very fast, to take persons and things about."

"Is that the thing you gave them for money?" I asked, pointing at the exhibit air-car.

"Yes, but it was new then. It has been made broken by things from guns now."

"What money did they give you for the machine?"

"One hundred pesos."

That started another uproar. There wasn't a soul in that courtroom who didn't know that five thousand pesos would have been a give-away bargain price for that car.

"Mr. Ambassador," one of the associate judges interrupted. "I used to be in the used-car business. Am I expected to believe that this ... this being ... sold that air-car for a hundred pesos?"

"Here's a notarized copy of the bill of sale, from the office of the Vehicles Registration Bureau," I said. "I introduce it as evidence."

There was a disturbance at the back of the room, and then the z'Srauff Ambassador, Gglafrr Ddespttann Vuvuvu, came stalking down the aisle, followed by a couple of Rangers and two of his attachés. He came forward and addressed the court.

"May you be happy, sir, but I am in here so quickly not because I have desire to make noise, but because it is only short time since it got in my knowledge that one of my persons is in this place. I am here to be of help to him that he not get in trouble, and to be of help to you. The name for what I am to do in this place is not part of my knowledge. Please say it for me."

"You are a friend of the court," Judge Nelson told him. "An *amicus curiae*."

"You make me happy. Please go on; I have no desire to put stop to what you do in this place."

"From what person did you get this machine that you gave to these persons for one hundred pesos?" I asked.

Gglafrr immediately began barking and snarling and yelping at my witness. The drygoods importer looked startled, and Judge Nelson banged with his gavel.

"That's enough of that! There'll be nothing spoken in this court but English, except through an interpreter!"

"Yow! I am sad that what I did was not right," the z'Srauff Ambassador replied contritely. "But my person here has not as part of his knowledge that you will make him say what may put him in trouble."

Nelson nodded in agreement.

"You are right: this person who is here has no need to make answer to any question if it may put him in trouble or make him seem less than he is."

"I will not make answer," the witness said.

"No further questions."

I turned to Goodham, and then to Sidney; they had no questions, either. I handed another slip of paper to the bailiff, and another z'Srauff, named Bbrarkk Jjoknyyegg Kekeke took the stand.

He put into this world things for small persons to make amusement with; he took out of this world meat and leather. He had his house of business in

New Austin, and he pointed out the three Bonneys as persons in this place that he saw that he had seen before.

"And what business did you make with them?" I asked.

"I gave them for money a gun which sends out things of twenty-millimeters very fast, to make death or hurt come to men and animals and does destruction to machines and things."

"Is this the gun?" I showed it to him.

"It could be. The gun was made in my world; many guns like it are made there. I am certain that this is the very gun."

I had a notarized copy of a customs house bill in which the gun was described and specified by serial number. I introduced it as evidence.

"How much money did these three persons give you for this gun?" I asked.

"Five pesos."

"The customs appraisal on this gun is six hundred pesos," I mentioned.

Immediately, Ambassador Vuvuvu was on his feet. "My person here has not as part of his knowledge that he may put himself in trouble by what he says to answer these questions."

That put a stop to that. Bbrarkk Jjoknyyegg Kekeke immediately took refuge in refusal to answer on grounds of self-incrimination.

"That is all, Your Honor," I said, "And now," I continued, when the witness had left the stand, "I have something further to present to the court, speaking both as *amicus curiae* and as Ambassador of the Solar League. This court cannot convict the three men who are here on trial. These men should have never been brought to trial in this court: it has no jurisdiction over this case. This was a simple case of first-degree murder, by hired assassins, committed against the Ambassador of one government at the instigation of another, not an act of political protest within the meaning of New Texan law."

There was a brief silence; both the court and the spectators were stunned, and most stunned of all were the three Bonney brothers, who had been watching, fear–sick, while I had been putting a rope around their necks. The uproar from the rear of the courtroom gave Judge Nelson a needed minute or so to collect his thoughts. After he had gotten order restored, he turned to me, grim–faced.

"Ambassador Silk, will you please elaborate on the extraordinary statement you have just made," he invited, as though every word had sharp corners that were sticking in his throat.

"Gladly, Your Honor." My words, too, were gouging and scraping my throat as they came out; I could feel my knees getting absurdly weak, and my mouth tasted as though I had an old copper penny in it.

"As I understand it, the laws of New Texas do not extend their ordinary protection to persons engaged in the practice of politics. An act of personal injury against a politician is considered criminal only to the extent that the politician injured has not, by his public acts, deserved the degree of severity with which he has been injured, and the Court of Political Justice is established for the purpose of determining whether or not there has been such an excess of severity in the treatment meted out by the accused to the injured or deceased politician. This gives rise, of course, to some interesting practices; for instance, what is at law a trial of the accused is, in substance, a trial of his victim. But in any case tried in this court, the accused must be a person who has injured or killed a man who is definable as a practicing politician under the government of New Texas.

"Speaking for my government, I must deny that these men should have been tried in this court for the murder of Silas Cumshaw. To do otherwise would establish the principle and precedent that our Ambassador, or any other Ambassador here, is a practicing politician under—mark that well, Your Honor—under the laws and government of New Texas. This would not only make of any Ambassador a permissable target for any marksman who happened to disapprove of the policies of another government, but more serious, it would place the Ambassador and his government in a subordinate position relative to the government of New Texas. This the government of the Solar League simply cannot tolerate, for reasons which it would be insulting to the intelligence of this court to enumerate."

"Mr. Silk," Judge Nelson said gravely. "This court takes full cognizance of the force of your arguments. However, I'd like to know why you permitted this trial to run to this length before entering this objection. Surely you could have made clear the position of your government at the beginning of this trial."

"Your Honor," I said, "had I done so, these defendants would have been released, and the facts behind their crime would have never come to light. I grant that the important function of this court is to determine questions of relative guilt and innocence. We must not lose sight, however, of the fact that the primary function of any court is to determine the truth, and only by the process of the trial of these depraved murderers–for–hire could the real author of the crime be uncovered.

"This was important, both for the government of the Solar League and the government of New Texas. My government now knows who procured the death of Silas Cumshaw, and we will take appropriate action. The government of New Texas has now had spelled out, in letters anyone can read, the fact that this beautiful planet is in truth a *battleground*. Awareness of this may save New Texas from being the scene of a larger and more destructive battle. New Texas also knows who are its enemies, and who can be counted upon to stand as its friends."

"Yes, Mr. Silk. Mr. Vuvuvu, I haven't heard any comment from you.... No comment? Well, we'll have to close the court, to consider this phase of the question."

The black screen slid up, for the second time during the trial. There was silence for a moment, and then the room became a bubbling pot of sound. At least six fights broke out among the spectators within three minutes; the Rangers and court bailiffs were busy restoring order.

Gail Hickock, who had been sitting on the front row of the spectators' seats, came running up while I was still receiving the congratulations of my fellow diplomats.

"Stephen! How *could* you?" she demanded. "You know what you've done? You've gotten those murdering snakes turned loose!"

Andrew Jackson Hickock left the prosecution table and approached.

"Mr. Silk! You've just secured the freedom of three men who murdered one of my best friends!"

"Colonel Hickock, I believe I knew Silas Cumshaw before you did. He was one of my instructors at Dumbarton Oaks, and I have always had the deepest respect and admiration for him. But he taught me one thing, which you seem to have forgotten since you expatriated yourself—that in the Diplomatic Service, personal feelings don't count. The only thing of importance is the advancement of the policies of the Solar League."

"Silas and I were attachés together, at the old Embassy at Drammool, on Altair II," Colonel Hickock said. What else he might have said was lost in the sudden exclamation as the black screen slid down. In front of Judge Nelson, I saw, there were three pistol–belts, and three pairs of automatics.

"Switchblade Joe Bonney, Jack–High Abe Bonney, Turkey–Buzzard Tom Bonney, together with your counsel, approach the court and hear the verdict," Judge Nelson said.

The three defendants and their lawyer rose. The Bonneys were swaggering and laughing, but for a lawyer whose clients had just emerged from the shadow of the gallows, Sidney was looking remarkably unhappy. He probably had imagination enough to see what would be waiting for him outside.

"It pains me inexpressibly," Judge Nelson said, "to inform you three that this court cannot convict you of the cowardly murder of that learned and honorable old man, Silas Cumshaw, nor can you be brought to trial in any other court on New Texas again for that dastardly crime. Here are your weapons, which must be returned to you. Sort them out yourselves, because I won't dirty my fingers on them. And may you regret and feel shame for your despicable act as long as you live, which I hope won't be more than a few hours."

With that, he used the end of his gavel to push the three belts off the bench and onto the floor at the Bonneys' feet. They stood laughing at him for a few moments, then stopped, picked the belts up, drew the pistols to check magazines and chambers, and then began slapping each others' backs and shouting jubilant congratulations at one another. Sidney's two assistants and some of his friends came up and began pumping Sidney's hands.

"There!" Gail flung at me. "Now look at your masterpiece! Why don't you go up and congratulate him, too?"

And with that, she slapped me across the face. It hurt like the devil; she was a lot stronger than I'd expected.

"In about two minutes," I told her, "you can apologize to me for that, or weep over my corpse. Right now, though, you'd better be getting behind something solid."

Chapter XI

I turned and stepped forward to confront the Bonneys, mentally thanking Gail. Up until she'd slapped me, I'd been weak–kneed and dry–mouthed with what I had to do. Now I was just plain angry, and I found that I was thinking a lot more clearly. Jack–High Bonney's wounded left shoulder, I knew, wouldn't keep him from using his gun hand, but his shoulder muscles would be stiff enough to slow his draw. I'd intended saving him until I'd dealt with his brothers. Now, I remembered how he'd gotten that wound in the first place: he'd been the one who'd used the auto–rifle, out at the Hickock ranch. So I changed my plans and moved him up to top priority.

"Hold it!" I yelled at them. "You've been cleared of killing a politician, but you still have killing a Solar League Ambassador to answer for. Now get your hands full of guns, if you don't want to die with them empty!"

The crowd of sympathizers and felicitators simply exploded away from the Bonney brothers. Out of the corner of my eye, I saw Sidney and a fat, blowsy woman with brass–colored hair as they both tried to dive under the friends–of–the–court table at the same place. The Bonney brothers simply stood and stared at me, for an instant, unbelievingly, as I got my thumbs on the release–studs of my belt. Judge Nelson's gavel was hammering, and he was shouting:

"Court–of–Political–Justice–Confederate–Continent–of–New–Texas–is– herewith– adjourned–reconvene–0900–tomorrow. *Hit the floor!*"

"Damn! He means it!" Switchblade Joe Bonney exclaimed.

Then they all reached for their guns. They were still reaching when I pressed the studs and the Krupp–Tattas popped up into my hands, and I swung up my right–hand gun and shot Jack–High through the head. After that, I just let my subconscious take over. I saw gun flames jump out at me from the Bonneys' weapons, and I felt my own pistols leap and writhe in my hands, but I don't believe I was aware of hearing the shots, not even from my own weapons. The whole thing probably lasted five seconds, but it seemed like twenty minutes to me. Then there was nobody shooting at me, and nobody for me to shoot at; the big room was silent, and I was aware that Judge Nelson and his eight associates were rising cautiously from behind the bench.

I holstered my left–hand gun, removed and replaced the magazine of the right–hand gun, then holstered it and reloaded the other one. Hoddy Ringo and Francisco Parros and Commander Stonehenge were on their feet, their pistols drawn, covering the spectators' seats. Colonel Hickock had also drawn a pistol and he was covering Sidney with it, occasionally moving the muzzle to the left to include the z'Srauff Ambassador and his two attachés.

By this time, Nelson and the other eight judges were in their seats, trying to look calm and judicial.

"Your Honor," I said, "I fully realize that no judge likes to have his court turned into a shooting gallery. I can assure you, however, that my action here was not the result of any lack of respect for this court. It was pure necessity. Your Honor can see that: my government could not permit this crime against its Ambassador to pass unpunished."

Judge Nelson nodded solemnly. "Court was adjourned when this little incident happened, Mr. Silk," he said.

He leaned forward and looked to where the three Bonney brothers were making a mess of blood on the floor. "I trust that nobody will construe my unofficial and personal comments here as establishing any legal precedent, and I wouldn't like to see this sort of thing become customary ... but ... you did that all by yourself, with those little beanshooters?...Not bad, not bad at all, Mr. Silk."

I thanked him, then turned to the z'Srauff Ambassador. I didn't bother putting my remarks into Basic. He understood, as well as I did, what I was saying.

"Look, Fido," I told him, "my government is quite well aware of the source from which the orders for the murder of my predecessor came. These men I just killed were only the tools.

"We're going to get the brains behind them, if we have to send every warship we own into the z'Srauff star–cluster and devastate every planet in it. We don't let dogs snap at us. And when they do, we don't kick them, we shoot them!"

That, of course, was not exactly striped–pants diplomatic language. I wondered, for a moment, what Norman Gazarian, the protocol man, would think if he heard an Ambassador calling another Ambassador Fido.

But it seemed to be the kind of language that Mr. Vuvuvu understood. He skinned back his upper lip at me and began snarling and growling. Then he turned on his hind paws and padded angrily down the aisle away from the front of the courtroom.

The spectators around him and above him began barking, baying, yelping at him: "Tie a can to his tail!" "Git for home, Bruno!"

Then somebody yelled, "Hey, look! Even his wrist watch is blushing!"

That was perfectly true. Mr. Gglafrr Ddespttann Vuvuvu's watch–face, normally white, was now glowing a bright ruby–red.

I looked at Stonehenge and found him looking at me. It would be full dark in four or five hours; there ought to be something spectacular to see in the cloudless skies of Capella IV tonight.

Fleet Admiral Sir Rodney Tregaskis would see to that.

>_FROM REPORT
>OF SPACE–COMMANDER STONEHENGE
>TO SECRETARY OF AGGRESSION, KLÜNG:

… so the measures considered by yourself and Secretary of State Ghopal Singh and Security Coördinator Natalenko, as transmitted to me by Mr. Hoddy Ringo, were not, I am glad to say, needed. Ambassador Silk, alive, handled the thing much better than Ambassador Silk, dead, could possibly have.

…to confirm Sir Rodney Tregaskis' report from the tales of the few survivors, the z'Srauff attack came as the Ambassador had expected. They dropped out of hyperspace about seventy light–minutes outside the Capella system, apparently in complete ignorance of the presence of our fleet.

...have learned the entire fleet consisted of about three hundred spaceships and reports reaching here indicate that no more than twenty got back to z'Srauff Cluster.

...naturally, the whole affair has had a profound influence, an influence to the benefit of the Solar League, on all shades of public opinion.

...as you properly assumed, Mr. Hoddy Ringo is no longer with us. When it became apparent that the Palme–Silk Annexation Treaty would be ratified here, Mr. Ringo immediately saw that his status of diplomatic immunity would automatically terminate. Accordingly, he left this system, embarking from New Austin for Alderbaran IX, mentioning, as he shook hands with me, something about a widow. By a curious coincidence, the richest branch bank in the city was held up by a lone bandit about half an hour before he boarded the space–ship...._

_FINAL MESSAGE
OF THE LAST SCLAR AMBASSADOR TO NEW
TEXAS
STEPHEN SILK

Copies of the Treaty of Annexation, duly ratified by the New Texas Legislature, herewith.

Please note that the guarantees of non–intervention in local political institutions are the very minimum which are acceptable to the people of New Texas. They are especially adamant that there will be no change in their peculiar methods of insuring that their elected and appointed public officials shall be responsible to the electorate.

DEPARTMENT ADDENDUM

After the ratification of the Palme–Silk treaty, Mr. Silk remained on New Texas, married the daughter of a local rancher there (see file on First Ambassador, Colonel Andrew Jackson Hickock) and is still active in politics on that planet, often in opposition to Solar League policies, which he seems to anticipate with an almost uncanny prescience.

Natalenko re–read the addendum, pursed his thick lips and sighed. There were so many ways he could be using Mr. Stephen Silk....

For example—he looked at the tri–di star–map, both usefully and beautifully decorating his walls—over there, where Hoddy Ringo had gone, near Alderbaran IX.

Those were twin planets, one apparently settled by the equivalent descendants of the Edwards and the other inhabited by the children of a Jukes–Kallikak union. Even the Solar League Ambassadors there had taken the viewpoints of the planets to whom they were accredited, instead of the all–embracing view which their training should have given them....

Curious problem ... and, how would Stephen Silk have handled it?

The Security Coördinator scrawled a note comprehensible only to himself....

www.ingramcontent.com/pod-product-compliance
Lightning Source LLC
Chambersburg PA
CBHW070114230526
45472CB00004B/1248